Marine Ecology and Ocean Sustainability

Ferdinand Jean C. Guendehou
Kpadédji Lambert Ayitchéhou

Marine Ecology and Ocean Sustainability

Challenges and Solutions

Springer

Ferdinand Jean C. Guendehou
Applied Ecology Laboratory
University of Abomey-Calavi
Cotonou, Benin

Kpadédji Lambert Ayitchéhou
Laboratory of Water and Environmental
Sciences and Techniques
University of Abomey-Calavi
Cotonou, Benin

ISBN 978-3-032-32755-0 ISBN 978-3-032-32756-7 (eBook)
https://doi.org/10.1007/978-3-032-32756-7

This Springer imprint is published by the registered company Springer Nature Switzerland AG
The registered company address is: Gewerbestrasse 11, 6330 Cham, Switzerland

Contents

About the Authors

Ferdinand Jean C. Guendehou is a researcher in applied ecology with a specialization in marine and coastal ecosystem management. He holds a Ph.D. in Environment Management from the University of Abomey. He is affiliated with the Applied Ecology Laboratory at the University of Abomey-Calavi, Benin (west Africa). His research focuses on the conservation of marine biodiversity, coastal resilience, climate change, and the integration of traditional ecological knowledge into modern environmental management frameworks. Dr. Guendehou has contributed to several scientific studies and international initiatives addressing ocean sustainability, ecosystem restoration, and community resource management. His work emphasizes interdisciplinary approaches that bridge science, policy, and local knowledge systems to improve the resilience of coastal ecosystems in West Africa and beyond. He is actively involved in research collaborations, capacity building, and policy-oriented projects aimed at promoting sustainable development and the conservation of marine and coastal environments.

Kpadédji Lambert Ayitchehou is a Benin-based expert in marine and coastal management, environmental planning, and climate resilience. He holds a Ph.D. in Geography and Environmental Management from the University of Abomey-Calavi and has over 28 years of professional experience across public administration, international development, and research. He serves as the Director General of Coastal Protection and Management and leads national programs addressing coastal erosion and marine spatial planning. His work integrates engineering, policy, and science to promote sustainable ocean governance. A researcher at the Laboratory of Water and Environmental Sciences, he has authored numerous peer-reviewed publications on marine ecosystems, climate change, and coastal infrastructure. He also teaches at the National School of Administration in Benin and mentors graduate students. His contributions advance resilient and inclusive coastal development.

Chapter 1
The Foundations of Marine Ecology

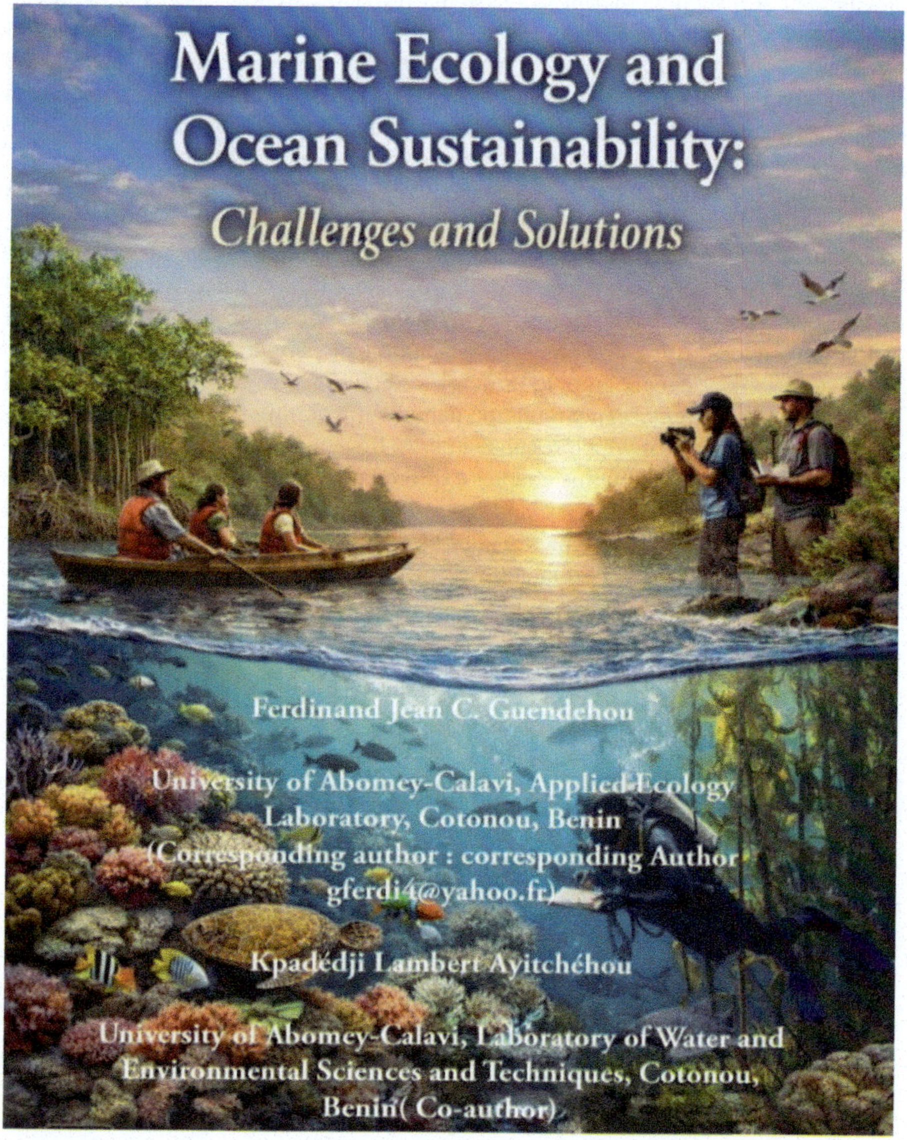

F. J. C. Guendehou, K. L. Ayitchéhou, *Marine Ecology and Ocean Sustainability*, https://doi.org/10.1007/978-3-032-32756-7_1

This study contributes to the scientific understanding of marine ecological systems by synthesizing fundamental concepts related to ocean structure, biodiversity, and oceanographic processes. By integrating ecological, physical, and chemical perspectives, the study highlights the interconnected nature of marine systems and emphasizes their critical role in global climate regulation, biodiversity conservation, and human well-being. Marine ecosystems provide essential services, including food resources for billions of people, coastal protection, carbon sequestration, and the maintenance of global biogeochemical cycles. Therefore, understanding the functioning and vulnerability of these systems is crucial for informing conservation strategies, sustainable fisheries management, and policy frameworks aimed at preserving ocean health in the context of accelerating environmental change.

This chapter provides a comprehensive overview of the foundations of marine ecology. First, it introduces the concept, scope, and importance of marine ecology, emphasizing the ecological, economic, and societal value of marine ecosystems. This chapter then describes the spatial organization of the ocean through horizontal and vertical zonation, explaining how environmental gradients such as depth, light availability, temperature, and nutrient distribution shape marine biodiversity. It further examines key oceanographic principles, including the physical and chemical properties of seawater, ocean circulation patterns, upwelling processes, and thermohaline circulation, all of which regulate energy flow, nutrient cycling, and climate–ocean interactions. Finally, this chapter explores the main marine habitats, such as coral reefs, mangroves, estuaries, and deep-sea environments, while discussing the roles of keystone species, trophic dynamics, and biodiversity hotspots in maintaining ecosystem stability and productivity. Together, these elements provide a scientific framework for understanding the functioning of marine ecosystems and highlight the urgent need for effective conservation and sustainable management of ocean resources.

1 Introduction to Marine Ecology

1.1 Definition and Scope

Marine ecology is the scientific study of marine life and the interactions of organisms with each other and their environment in oceanic and coastal habitats [1]. As a branch of ecology, it focuses on understanding the complex relationships between marine organisms ranging from microscopic plankton to massive whales and the physical, chemical, and biological factors that influence their distribution, abundance, and behavior [2]. The marine environment covers more than 70% of the Earth's surface and contains various ecosystems, such as coral reefs, estuaries, mangroves, open oceans, and deep-sea habitats [3]. These ecosystems support a wide variety of life forms and provide critical services, including oxygen production, climate regulation, carbon storage, and food resources for billions of people [4]. The scope of marine ecology is vast and multidisciplinary, integrating biology, oceanography, chemistry, geology, and environmental science [5]. It encompasses various levels of biological organization from genes and cells to populations, communities, and ecosystems [1]. Marine ecology includes several key areas. Population and community ecology involves studying the dynamics of marine populations, species interactions (such as predation, competition, and symbiosis), and the structure and function of marine communities [6]. Ecosystem ecology focuses on the flow of energy and the cycling of nutrients through marine ecosystems, assessing how physical processes such as currents, tides, and temperature gradients affect biological activity [6]. According to conservation and management, marine ecology plays a critical role in understanding human impacts such as overfishing, pollution, habitat destruction, and climate change in marine systems [7] and it provides the scientific basis for the conservation of biodiversity and the sustainable management of marine resources [8]. Applied marine ecology includes the practical application of ecological principles to solve marine environmental problems [9]. It involves habitat restoration, marine protected area (MPA) design, impact assessments, and the development of strategies to mitigate the effects of global change [10]. Behavioral and physiological ecology examines how marine organisms adapt to their environments through physiological mechanisms and behavioral strategies, especially under changing environmental conditions [11]. Marine ecology is essential not only to advance scientific knowledge but also to ensure the health and resilience of marine ecosystems in the face of escalating anthropogenic pressures [12]. As our dependence on marine resources increases, so does the importance of marine ecological research in guiding responsible stewardship of the oceans [13].

1.2 Importance of Marine Ecosystems

Marine ecosystems are among the most vital components of the Earth's biosphere. Covering more than 70% of the planet's surface, the oceans and seas play a fundamental role in supporting life on Earth [14]. From microscopic phytoplankton that form the base of the marine food web to vast coral reefs and deep-sea ecosystems, marine environments provide critical ecological, economic, and cultural services [15]. Understanding the importance of marine ecosystems is a fundamental aspect of marine ecology, a field dedicated to studying the relationships between marine organisms and their physical and biological environments [7]. Speaking of *biodiversity hotspots and habitat variety*, marine ecosystems are incredibly diverse. They range from coastal environments such as estuaries, salt marshes, mangroves, and coral reefs to open ocean and deep-sea habitats [3]. These ecosystems support an extraordinary variety of species, many of which are endemic and not found anywhere else [16]. Coral reefs, often called "rainforests of the sea," are particularly rich in biodiversity, providing habitat for approximately 25% of all known marine species despite covering less than 1% of the ocean floor [17]. This biodiversity is crucial for ecosystem stability, resilience, and productivity [18]. A diverse marine ecosystem can better withstand environmental stress and recover from disturbances, such as storms, pollution, or climate change [19]. Regarding *climate regulation and carbon sequestration*, marine ecosystems play a central role in the Earth's climate [20]. Oceans absorb approximately 90% of excess heat generated by greenhouse gas emissions and roughly 25–30% of atmospheric carbon dioxide [21]. Phytoplanktons, through the process of photosynthesis, contribute significantly to global oxygen production estimated at more than 50% of the oxygen supply [22]. Furthermore, blue carbon ecosystems such as mangroves, seagrasses, and salt marshes are particularly effective in sequestering carbon in their soils and biomass for long periods [23]. This makes them critical allies in mitigating climate change and its impacts [24]. About *food security and livelihoods*, marine ecosystems are a primary source of food and income for billions of people around the world. Fisheries and aquaculture provide more than 3 billion people with at least 20% of their animal protein intake [25]. Coastal communities, especially in developing countries, are highly dependent on marine resources for food and economic activity. In addition to commercial and subsistence fishing, marine ecosystems support tourism, transportation, and other maritime industries, which contribute significantly to global and national economies [26]. Coral reefs, for example, attract millions of tourists annually, generate billions in revenue, and support local jobs [27]. With *natural defense and disaster mitigation, coastal* marine ecosystems such as mangroves, coral reefs, and salt marshes act as natural barriers that protect shorelines from erosion, storm surges, and tsunamis. By dissipating wave energy, these ecosystems reduce the impact of extreme weather events on coastal infrastructure and human populations [28]. Degradation or loss of these protective ecosystems increases the vulnerability of coastal areas to climate-induced disasters, which are expected to become more frequent and intense due to global warming [29]. In view of *biochemical and genetic resources*, marine organisms produce a wide array of bioactive compounds with potential applications in medicine, biotechnology, and industry. Marine-derived substances have already led to the development of new drugs for the treatment of cancer, pain, bacterial infections, and inflammatory diseases [30]. The unique biochemical and genetic diversity found in marine species offers immense potential for future discoveries in pharmaceuticals, cosmetics, and other fields [31]. Furthermore, marine ecosystems contribute to the cycling of nutrients, especially nitrogen and phosphorus, which are essential for global biogeochemical processes and agricultural productivity [32]. Regarding cultural, recreational, and aesthetic value, marine environments hold significant cultural, spiritual, and recreational value for many coastal and indigenous communities [33]. Oceans and seas have inspired countless myths, artworks, and traditions throughout human history. Activities such as diving, surfing, whale watching, and sailing also contribute to physical well-being, mental health, and cultural identity [34]. The aesthetic and intrinsic value of marine ecosystems further reinforces the need to conserve them, not just for their utility, but also for their beauty and their role in enriching human life [35]. With regard to scientific and educational importance, marine ecosystems are natural laboratories for studying ecological and evolutionary processes. They help scientists understand how life adapts to extreme environments, how ecosystems respond to stress, and how species interact in complex networks [36]. Marine ecology provides critical information on broader ecological principles and supports education and awareness of environmental stewardship [37]. About *global interconnectivity*, the oceans are interconnected through global currents, climate systems, and the movement of species. Changes in one marine region, such as coral bleaching in the Pacific or melting ice in the Arctic, can have ripple effects around the globe [38]. This connectivity underscores the importance of international cooperation in marine conservation and sustainable management [39]. *In conclusion*, marine ecosystems are essential to life on Earth. They support biodiversity, regulate climate, sustain economies, protect coastlines, provide food and medicine, and enrich human culture [40]. However, these ecosystems are increasingly threatened by human activities, including pollution, overfishing, habitat destruction, and climate change. Marine ecology plays a vital role in understanding these systems and guiding conservation and management strategies [41]. Recognizing the importance of marine ecosystems is not merely an academic exercise, it is a prerequisite for ensuring a sustainable and resilient future for both marine life and humanity [42]. As stewards of the oceans of the planet, it is our collective responsibility to protect and preserve these invaluable ecosystems for generations to come [13].

1.3 Overview of Ocean Zones and Biodiversity

Marine ecology, the study of the interactions among marine organisms and their environment, requires a fundamental understanding of the physical structure of the ocean [7]. The marine environment is vast and varied, comprising multiple zones that differ in depth, light availability, temperature, pressure, and nutrient content [43]. Each of these ocean zones supports unique biological communities, contributing to the incredible biodiversity found in marine ecosystems. Understanding the structure of ocean zones and the distribution of biodiversity within them is critical to the study of marine ecology. It helps ecologists investigate species adaptations, ecological interactions, and the environmental factors that shape life beneath the waves [1].

1.4 Ocean Zones: Vertical and Horizontal Divisions

Marine ecologists typically classify the ocean into vertical (depth-based) and horizontal (distance from shore) zones to better understand how different environmental conditions influence marine life [44]. According to *horizontal zones (based on distance from shore)* the ocean is divided into horizontal zones based on distance from shore: the littoral zone, the neritic zone, and the oceanic zone [45]. The littoral zone (or intertidal zone) lies between high and low tide marks, where organisms such as barnacles, sea stars, and crabs are adapted to withstand changing moisture, temperature, and salinity [46]. Extending from the low tide line to the edge of the continental shelf is the neritic zone, a sunlit area that supports diverse ecosystems such as coral reefs and kelp forests and is home to fish, invertebrates, marine mammals, and seabirds [47]. Beyond the continental shelf lies the oceanic zone, characterized by deep open waters with lower nutrient levels of nutrients but still supporting large migratory species, such as tuna, sharks, and whales [48]. Speaking of vertical zones (*based on depth and penetration of light*), the vertical zones, defined by depth and light availability, range from sunlit surfaces to the deepest trenches [49]. The epipelagic zone (0–200 m), or sunlight zone, receives ample light, allowing photosynthesis and supporting most marine life, including plankton, jellyfish, and large fish [50]. Below it, the mesopelagic zone (200–1000 m), or the twilight zone, is dimly lit and home to bioluminescent species, such as lanternfish and squid [51]. The bathypelagic zone (1000–4000 m), or the midnight zone, is pitch dark, cold, and under high pressure, inhabited by slow-metabolism creatures, such as anglerfish and gulper eels [52]. Deeper still, the abyssopelagic zone (4000–6000 m) has near freezing temperatures and sparse life adapted to extreme pressure, such as basket stars and sea cucumbers [53]. The deepest layer, the hadalpelagic zone (6000+ m), exists in ocean trenches and hosts highly specialized organisms, such as amphipods and extremophile bacteria [54].

1.5 Marine Biodiversity Across Zones

Marine biodiversity varies widely between ocean zones, shaped by physical conditions. Such as depth, light, temperature, and habitat type [55]. Coral reefs, found in warm, shallow neritic waters, are among the most diverse ecosystems on Earth, supporting thousands of species but are highly sensitive to environmental changes [56]. Mangrove forests, located in tropical intertidal zones, serve as nurserylands for marine life and protect coastlines, hosting terrestrial and aquatic species [57]. Seagrass beds, composed of flowering plants in shallow waters, provide shelter and food for fish, turtles, and invertebrates while stabilizing sediments and improving water quality [58]. In the open ocean, large migratory animals such as whales, tuna, and sharks rely on plankton-based food webs [59]. Deep-sea ecosystems, such as hydrothermal vents and abyssal plains, support life through chemosynthesis, hosting unique organisms, such as giant tubeworms and extremophile bacteria [60]. Finally, the polar regions, although cold, are seasonally rich in life due to the increase in nutrients, supporting species, such as krill, seals, whales, penguins, and polar bears [61].

1.6 Importance of Zonal Biodiversity

Zonal marine biodiversity is essential for maintaining healthy ocean ecosystems. Each zone hosts species uniquely adapted to its conditions, collectively supporting vital functions, such as food webs, nutrient cycling, and oxygen production [62].

This diversity improves the resilience of the ecosystem to threats, such as climate change and pollution, while also providing critical services to humans, including fishing, coastal protection, and overall ocean productivity [63].

1.7 Threats to Marine Biodiversity

Marine biodiversity is threatened significantly by human activities. Overfishing disrupts food webs and depletes fish stocks, while plastic pollution, oil spills, and agricultural runoff harm marine organisms [64]. Climate change contributes to ocean warming, acidification, and coral bleaching, and habitat destruction through coastal development and dredging further degrades critical ecosystems [65]. To protect marine life, conservation strategies such as marine protected areas (MPA), sustainable fishing, and international cooperation are crucial for preserving ocean health and biodiversity [66].

2 Oceanographic Principles

2.1 Physical and Chemical Properties of Seawater

The physics of seawater is central to oceanography, as it governs the behavior and movement of the oceans [67]. Seawater is mainly composed of water and dissolved salts, with physical properties such as temperature, salinity, density, pressure, light, and sound transmission playing key roles [68]. Temperature and salinity vary between different regions and depths, directly affecting the density of seawater [69]. Denser water sinks, while less dense water rises, driving vertical movement and forming the basis of thermohaline circulation, which distributes heat and nutrients across the world [70]. The pressure increases with depth, compressing the water slightly and influencing the behavior of marine organisms and chemical reactions [71]. Light penetrates only the upper layers of the ocean (typically the top 100–200 m), limiting photosynthesis to the photic zone. Below this, the ocean is dark, and biological activity changes accordingly. Sound travels faster and farther in seawater than in air, making it a valuable tool for marine navigation, communication, and research [72]. Seawater also has specific thermal properties, such as high heat capacity, allowing it to absorb and store large amounts of heat, which plays a vital role in climate regulation [73]. Understanding the physical properties of seawater helps oceanographers study currents, waves, tides, and the global climate system, making it essential for marine science, navigation, and environmental management [74]. *The chemical properties of seawater* are vital to understanding marine environments and the global ocean system [75]. Seawater is a complex chemical solution composed mainly of water (H_2O) and dissolved salts, gases, and organic matter [76]. Its salinity, typically around 35 parts per thousand (ppt), is determined by the concentration of dissolved salts, mainly sodium (Na^+) and chloride (Cl). Other key ions include sulfate (SO42), magnesium (Mg^{2+}), calcium (Ca^{2+}), and potassium (K^+). Despite regional variations in salinity, the relative proportions of these ions remain constant, known as the principle of constant proportions [77]. Seawater also contains essential nutrients, such as nitrate (NO3), phosphate (PO_4^{3-}), and silicate (SiO44), which support marine life, particularly phytoplankton. These nutrients cycle through biological activity and are influenced by upwelling, mixing, and biological decay. Dissolved gases are another important chemical component [78]. Oxygen (O_2) is critical for marine respiration, while carbon dioxide (CO_2) interacts with water to form carbonic acid, bicarbonate, and carbonate ions, forming the carbonate buffer system [79]. This system helps regulate the pH of seawater, typically between 7.5 and 8.4 [80]. Trace elements and organic compounds also contribute to the chemistry of seawater, affecting biological and chemical interactions. Understanding the chemical properties of seawater is essential for studying ocean productivity, acidification, pollution, and biogeochemical cycles [76].

2.2 *Ocean Currents, Upwelling, and Thermohaline Circulation*

Ocean currents are large-scale movements of seawater that play a crucial role in regulating the climate of the Earth, distributing heat, and supporting marine ecosystems [81]. They are driven by a combination of forces, including wind patterns, Earth rotation (Coriolis effect), differences in water density due to temperature and salinity (thermohaline circulation), and the shape of ocean basins [82]. Surface currents, which occur in the upper 400 m of the ocean, are primarily driven by wind and form large circular systems called gyres [83]. These gyres dominate the main ocean basins: the North and South Atlantic, North and South Pacific, and the Indian Ocean. The Gulf Stream in the North Atlantic and the Kuroshio Current in the North Pacific are examples of powerful warm currents that transport heat from the tropics toward the poles [84]. Deep ocean currents, part of the global conveyor belt or thermohaline circulation, are driven by differences in water density [70]. Cold salty water sinks in polar regions and flows slowly along the ocean floor toward the equator, where it gradually rises and warms [70]. This circulation helps regulate global climate by transporting heat and carbon between the surface and deep ocean [85]. Currents also influence weather patterns, nutrient distribution, and marine life migration [86]. Upwelling currents, which bring deep-water rich in nutrients to the surface, support highly productive fisheries [87]. Understanding ocean currents is essential for navigation, climate prediction, and environmental management [88].

Upwelling is a vital oceanographic process in which deep, cold, and nutrient-rich water rises to the ocean surface, replacing warmer surface water. This process plays a key role in supporting marine ecosystems and influencing the global climate. Upwelling occurs primarily along coastlines, the equator, and areas where wind patterns and ocean currents interact in specific ways. Coastal upwelling is usually caused by wind blowing parallel to the shore [89]. Due to the Coriolis effect, surface water is pushed away from the coast and deeper water moves upward to replace it [90]. Major coastal upwelling zones occur off the coasts of California, Peru,

Northwest Africa, and Namibia. These regions are among the most biologically productive in the world, supporting large populations of plankton, fish, and marine mammals [91]. Equatorial upwelling occurs when trade winds push surface waters away from the equator, allowing deeper water to rise. This process is especially significant in the Pacific and Atlantic Oceans and is crucial to maintaining equatorial marine ecosystems [92]. Upwelled water is rich in nutrients, such as nitrate and phosphate, which fuel the growth of the base of the marine food web. As a result, upwelling zones are critical to global fisheries and biodiversity [93]. Furthermore, upwelling influences climate by affecting sea surface temperatures and carbon dioxide exchange. Understanding upwelling is essential for managing marine resources, predicting fish yields, and monitoring climate change [94].

Thermohaline circulation is a global system of deep ocean currents driven by differences in water temperature (thermo) and salinity (haline), which together determine the density of seawater [95]. This circulation plays a critical role in the regulation of the climate of the Earth, transporting heat, carbon, oxygen, and nutrients across the world's oceans [96]. The process begins in high-latitude regions, particularly the North Atlantic and around Antarctica, where cold temperatures and high salinity make surface water dense enough to sink to the deep ocean [69]. This sinking water initiates deep currents that slowly flow along the ocean floor toward the equator and into other ocean basins [97]. Eventually, this deep water gradually rises back to the surface through upwelling in various regions, completing the global loop known as the "global conveyor belt" [98]. Thermohaline circulation operates on long timescales, taking hundreds to thousands of years to complete a cycle [70]. It helps to moderate global climate by moving warm water from the tropics to the poles and bringing cold water back toward the equator [99]. Additionally, it supports marine ecosystems by circulating oxygen-rich surface water to the deep sea and returning nutrient-rich deep water to the surface. Any disruption to this system, such as increased freshwater input from melting ice, can slow or alter circulation patterns, potentially impacting global climate and ocean health [100]. Understanding thermohaline circulation is essential to predict long-term climate change and ocean behavior [70].

2.3 Climate-Ocean Interactions

Climate–ocean interactions are fundamental to understanding the climate system, as the ocean and atmosphere continuously exchange heat, moisture, gases, and momentum [101]. The ocean acts as a vast heat reservoir, absorbing solar radiation and redistributing it through currents, helping to moderate global temperatures. This interaction influences weather patterns, climate variability, and long-term climate change [38]. One of the most significant examples is the El Niño-Southern Oscillation (ENSO), a periodic shift in ocean temperatures and atmospheric conditions in the equatorial Pacific [102]. During El Niño events, warmer-than-average sea surface temperatures disrupt global weather patterns, causing droughts, floods, and storms in various regions. On the contrary, La Niña events bring cooler waters and opposite climate effects [103]. These cycles highlight the powerful influence of ocean temperatures on global atmospheric circulation [104]. Oceans also absorb about 90% of the excess heat from greenhouse gas emissions and roughly 25% of atmospheric carbon dioxide, playing a key role in buffering climate change [105]. However, this leads to ocean warming, acidification, and sea-level rise, which in turn affect marine ecosystems, weather patterns, and coastal communities [106]. Understanding the interactions between the climate and oceans is critical for predicting climate variability, managing natural disasters, and planning for long-term environmental change [107]. It requires an integrated study of oceanography, meteorology, and climate science to forecast future conditions and adapt to global challenges [108].

3 Marine Biodiversity and Habitats

3.1 Coral Reefs, Mangroves, Estuaries, and Deep-Sea Environments

Coral reefs are among the most diverse and valuable marine ecosystems on Earth, often called the "rainforests of the sea" [109]. Found in shallow, warm tropical waters, they are formed by coral polyp colonies that build calcium carbonate skeletons over thousands of years [110]. Coral reefs support more than 25% of all marine species, including fish, mollusks, crustaceans, and algae, despite covering less than 1% of the ocean floor [111]. They provide vital ecosystem services including coastal protection from storms, habitat for marine life, food security, and economic benefits through tourism and fishing [112]. Coral reef biodiversity depends on clean, clear, warm water with stable salinity and low nutrient levels. However, these ecosystems are increasingly threatened by climate change, ocean acidification, pollution, overfishing, and destructive practices, such as blast fishing [113]. Coral bleaching caused by prolonged thermal stress has decimated reef systems around the world [114]. Conservation efforts focus on marine protected areas, coral gardening, sustainable fisheries, and global actions to reduce greenhouse gas emissions [66].

Mangroves are intertidal forests that are found along tropical and subtropical coastlines. Characterized by salt-tolerant trees and shrubs, mangroves are uniquely adapted to brackish water, oxygen-poor soils, and tidal fluctuations. These habitats support rich biodiversity, including fish, crustaceans, birds, and reptiles. They act as critical nurseries for many marine species and provide feeding grounds for migratory birds [115]. Beyond biodiversity, mangroves provide key ecosystem services: they stabilize coastlines, prevent erosion, sequester significant amounts of carbon, and serve as natural buffers against storm surges and tsunamis [116]. Despite their value, mangroves are under threat from aquaculture (especially shrimp farming), coastal development, pollution, and rising sea levels [117]. Conservation strategies include reforestation, community management, legal protections, and integration into climate change mitigation policies due to their carbon storage potential [118].

Estuaries are dynamic environments where freshwater from rivers mixes with saltwater from the sea, creating nutrient-rich brackish waters. They include bays, lagoons, and river mouths, supporting highly productive ecosystems that support diverse biological communities [119]. Estuaries serve as crucial breeding, spawning, and nursery grounds for many fish and shellfish, making them vital to commercial and subsistence fisheries. Estuarine biodiversity includes fish, crustaceans, mollusks, birds, and plants, such as saltmarsh grasses [120]. Their ecological functions include water filtration, carbon storage, and shoreline stabilization. However, estuaries are among the most threatened marine environments due to land reclamation, urbanization, agricultural runoff, and industrial pollution [121]. Habitat restoration, improved watershed management, and pollution control are essential to maintaining estuarine health and productivity [122].

Deep-sea environments typically below 200 m encompass the largest and least explored ecosystems on Earth. These areas include abyssal plains, seamounts, hydrothermal vents, and deep ocean trenches [123]. Life persists in extreme conditions: cold temperatures, high pressure, and total darkness. Biodiversity is surprisingly rich, featuring specialized organisms, such as bioluminescent fish, deep-sea corals, tubeworms, and extremophiles [124]. Deep-sea ecosystems play a critical role in nutrient cycling, carbon sequestration, and the support of global biodiversity. Hydrothermal vents host unique life forms that are dependent on chemosynthesis instead of photosynthesis [125]. These environments are increasingly threatened by deep-sea mining, climate change, plastic pollution, and overfishing, particularly slow-growing species like orange roughy [126]. Since recovery is extremely slow due to low reproductive rates, precautionary and international regulation is key to protecting deep-sea biodiversity and function [127].

3.2 Keystone Species and Trophic Dynamics

Keystone species are organisms that play a critical role in maintaining the structure, function, and diversity of marine ecosystems. Although their abundances are often low, their impact on ecological processes is disproportionately large [128]. The removal of a keystone species can cause cascading effects throughout the ecosystem, altering the composition of the species and disrupting ecosystem services [129]. In marine habitats, keystone species regulate population dynamics, maintain habitat structure, and enable the coexistence of diverse organisms. The survival of keystone species is threatened by overfishing, climate change, pollution, habitat destruction, and invasive species [130]. Their protection is essential to preserve marine biodiversity and ensure ecosystem resilience. Conservation strategies include the creation of marine protected areas (MPA), the regulation of fisheries, and the restoration of degraded habitats [66]. By focusing on keystone species, we can protect entire ecosystems and maintain the critical services they provide to humanity and the planet [131]. *Trophic dynamics* describe how energy and nutrients flow through marine ecosystems, from primary producers to top predators and decomposers. This flow structures food webs and determines the productivity, stability, and biodiversity of the ecosystem [132]. Energy transfer between levels is inefficient, with only ~10% of energy moving from one level to the next [133]. This means that disturbances at lower levels (e.g., reduced phytoplankton due to warming oceans) can significantly affect the entire system [134]. *Trophic cascades* occur when changes in one trophic level cause ripple effects across others [135]. For instance, overfishing apex predators can lead to mesopredator release, where mid-level species overconsume herbivores, resulting in habitat degradation (e.g., coral decline due to unchecked algae growth) [136]. Bottom-up changes, such as nutrient pollution, can trigger algal blooms, hypoxia, and fish die-offs common in coastal eutrophication zones [137]. Climate change, habitat loss, and pollution disrupt the trophic dynamics by altering the distribution, abundance, and interactions of species [138]. Maintaining balanced food webs requires sustainable fishing, habitat protection, pollution control, and integrated ecosystem-based management [139]. Healthy trophic dynamics supports biodiversity, fisheries, carbon cycling, and ocean resilience, making their protection essential for marine and human well-being [140].

3.3 Biodiversity Hotspots and Conservation Value

Biodiversity hotspots are regions with exceptionally high levels of species richness and endemism that are also experiencing severe habitat loss [141]. First identified by the ecologist Norman Myers, a region must meet two criteria to qualify: it must contain at least 1500 species of endemic vascular plants and must have lost over 70% of its original natural vegetation [142]. Currently, 36 biodiversity hotspots have been recognized globally, covering just 2.3% of the land surface but containing more than 50% of endemic plant species and a large share of terrestrial vertebrates [141]. Hotspots are found in various biomes, including tropical rainforests, mountain ranges, Mediterranean-type ecosystems, and coral reefs. Examples include the Amazon Basin, Madagascar and the Indian Ocean Islands, the Himalayas, and the Indo-Burma region [143]. In the marine realm, equivalents such as the Coral Triangle and the Western Indian Ocean are rich in coral species, fish, and invertebrates, many of which are found nowhere else [144]. These areas are vital ecological centers, but they are also highly vulnerable to threats, such as deforestation, agricultural expansion, urban development, climate change, pollution, and invasive species [145]. Many hotspots are located in low- and middle-income countries where human populations depend heavily on natural resources, further complicating conservation efforts. Despite their small global footprint, biodiversity hotspots are essential for planetary health [146]. Protecting them ensures the survival of thousands of unique species, many of which are on the brink of extinction. Due to their high biological value and urgent conservation need, hotspots are a global priority for biodiversity protection efforts. The conservation value of biodiversity hotspots lies in their irreplaceable species, ecosystem services, and the ecological stability they support. These areas harbor a large share of endemic species, meaning their loss would result in global extinctions [147]. Beyond preserving wildlife, conserving hotspots ensures the continued functioning of natural systems that regulate climate, purify water, support pollination, prevent erosion, and store carbon [148]. Hotspots also provide substantial economic and cultural value. Many indigenous communities rely on these ecosystems for food, medicine, and cultural practices. Additionally, biodiversity supports industries, such as ecotourism, agriculture, and fisheries [149]. For example, mangroves and coral reefs in marine hotspots protect coastal communities from storms and rising sea level while sustaining fisheries. Given the rapid degradation of these ecosystems, strategic conservation in hotspots yields high returns [150]. Targeted protection can conserve large numbers of species with limited investment. Preserving biodiversity hotspots is critical not only for safeguarding rare species but also for maintaining the life support systems of our planet. Effective conservation in these regions supports both environmental and human well-being, making them a global priority in the fight against loss of biodiversity and climate change [151] (Figs. 1.1, 1.2 and 1.3).

Fig. 1.1 This photograph illustrates a dense mangrove ecosystem observed at Togbin, Benin (West Africa), during a scientific canoe survey conducted within the framework of biodiversity conservation. The vigorous vegetation and dense canopy cover reflect a relatively functional coastal habitat, playing a crucial role in shoreline stabilization, carbon sequestration, and protection against coastal erosion. The turbid waters indicate active sedimentary dynamics that promote nutrient exchange processes. This environment also constitutes an important ecological refuge for numerous aquatic and avian species, thereby highlighting the critical importance of its sustainable conservation and management. **Source Image**: Dr. Ferdinand GUENDEHOU (2026)

Fig. 1.2 This photograph taken in Togbin, Benin (West Africa), illustrates the successful natural regeneration of mangrove seedlings, promoted through the active involvement of local communities in natural resource conservation. The young plants exhibit good vegetative vigor, characterized by predominantly green and well-developed leaves, reflecting effective adaptation to the coastal environment. This regeneration dynamic contributes significantly to soil stabilization, the reduction of coastal erosion, and the preservation of aquatic habitats. **Source Image**: Dr. Kpadédji L. Ayitchehou (2026)

Fig. 1.3 Co-management of marine resources with the participation of local communities and scientists. Collaborative governance systems enhance the sustainability and resilience of marine ecosystems. **Source Image**: Dr. Ferdinand GUENDEHOU (2026)

References

1. Kaiser, M. J. (2011). *Marine ecology: Processes, systems, and impacts*. Oxford University Press.
2. Andersen, K. H., Berge, T., Gonçalves, R. J., Hartvig, M., Heuschele, J., Hylander, S., et al. (2016). Characteristic sizes of life in the oceans, from bacteria to whales. *Annual Review of Marine Science, 8*(1), 217–241.
3. El-Regal, M. A., & Satheesh, S. (2023). Biodiversity of marine ecosystems. In *Marine ecosystems: A unique source of valuable bioactive compounds* (Vol. 3, pp. 1–42). Bentham Books.
4. Brusseau, M. L. (2019). Ecosystems and ecosystem services. In *Environmental and pollution science* (pp. 89–102). Academic Press.
5. Munn, C., & Munn, C. B. (2019). *Marine microbiology: Ecology & applications*. CRC Press.
6. Bertness, M. D., Gaines, S. D., & Hay, M. E. (Eds.). (2001). *Marine community ecology* (p. 550). Sinauer Associates.
7. Mann, K. H., & Lazier, J. R. (2005). *Dynamics of marine ecosystems: Biological-physical interactions in the oceans*. John Wiley & Sons.
8. Castilla, J. C. (2000). Roles of experimental marine ecology in coastal management and conservation. *Journal of Experimental Marine Biology and Ecology, 250*(1–2), 3–21.
9. Peterson, C. H. (1993). Improvement of environmental impact analysis by application of principles derived from manipulative ecology: Lessons from coastal marine case histories. *Australian Journal of Ecology, 18*(1), 21–52.
10. McLeod, E., Salm, R., Green, A., & Almany, J. (2009). Designing marine protected area networks to address the impacts of climate change. *Frontiers in Ecology and the Environment, 7*(7), 362–370.

11. Blewett, T. A., Binning, S. A., Weinrauch, A. M., Ivy, C. M., Rossi, G. S., Borowiec, B. G., et al. (2022). Physiological and behavioural strategies of aquatic animals living in fluctuating environments. *Journal of Experimental Biology, 225*(9), jeb242503.
12. Bulleri, F., Batten, S., Connell, S. D., Benedetti-Cecchi, L., Gibbons, M., Nugues, M. M., & Gribben, P. (2020). Human pressures and the emergence of novel marine ecosystems. In *Oceanography and marine biology* (pp. 456–535). CRC Press.
13. National Research Council, Division on Earth, Life Studies, Ocean Studies Board, & Committee on International Capacity-Building for the Protection, Sustainable Use of Oceans, & Coasts. (2008). *Increasing capacity for stewardship of oceans and coasts: A priority for the 21st century*. National Academies Press.
14. Chakraborty, S. K. (2023). Ocean ecosystem and its multidimensional eco-functionality and significance. In *The Palgrave handbook of global sustainability* (pp. 537–581). Springer International Publishing.
15. Thurber, A. R., Sweetman, A. K., Narayanaswamy, B. E., Jones, D. O., Ingels, J., & Hansman, R. L. (2014). Ecosystem function and services provided by the deep sea. *Biogeosciences, 11*(14), 3941–3963.
16. Mittermeier, R. A., Turner, W. R., Larsen, F. W., Brooks, T. M., & Gascon, C. (2011). Global biodiversity conservation: The critical role of hotspots. In *Biodiversity hotspots: Distribution and protection of conservation priority areas* (pp. 3–22). Springer.
17. Knowlton, N., Brainard, R. E., Fisher, R., Moews, M., Plaisance, L., & Caley, M. J. (2010). Coral reef biodiversity. In *Life in the world's oceans: Diversity distribution and abundance* (pp. 65–74). Wiley-Blackwell.
18. McClanahan, T., Polunin, N., & Done, T. (2002). Ecological states and the resilience of coral reefs. *Conservation Ecology, 6*(2).
19. O'Leary, J. K., Micheli, F., Airoldi, L., Boch, C., De Leo, G., Elahi, R., et al. (2017). The resilience of marine ecosystems to climatic disturbances. *Bioscience, 67*(3), 208–220.
20. Howard, J., Sutton-Grier, A., Herr, D., Kleypas, J., Landis, E., Mcleod, E., et al. (2017). Clarifying the role of coastal and marine systems in climate mitigation. *Frontiers in Ecology and the Environment, 15*(1), 42–50.
21. Abas, N., & Khan, N. (2014). Carbon conundrum, climate change, CO2 capture and consumptions. *Journal of CO2 Utilization, 8*, 39–48.
22. Sekerci, Y., & Petrovskii, S. (2018). Global warming can lead to depletion of oxygen by disrupting phytoplankton photosynthesis: A mathematical modelling approach. *Geosciences, 8*(6), 201.
23. Pham, T. D., Xia, J., Ha, N. T., Bui, D. T., Le, N. N., & Takeuchi, W. (2019). A review of remote sensing approaches for monitoring blue carbon ecosystems: Mangroves, seagrasses and salt marshes during 2010–2018. *Sensors, 19*(8), 1933.
24. Schaeffer-Novelli, Y., Soriano-Sierra, E. J., Vale, C. C. D., Bernini, E., Rovai, A. S., Pinheiro, M. A. A., et al. (2016). Climate changes in mangrove forests and salt marshes. *Brazilian Journal of Oceanography, 64*(spe2), 37–52.
25. Pauly, D., Watson, R., & Alder, J. (2005). Global trends in world fisheries: Impacts on marine ecosystems and food security. *Philosophical Transactions of the Royal Society B: Biological Sciences, 360*(1453), 5–12.
26. Ferrol-Schulte, D., Wolff, M., Ferse, S., & Glaser, M. (2013). Sustainable Livelihoods Approach in tropical coastal and marine social–ecological systems: A review. *Marine Policy, 42*, 253–258.
27. Spalding, M., Burke, L., Wood, S. A., Ashpole, J., Hutchison, J., & Zu Ermgassen, P. (2017). Mapping the global value and distribution of coral reef tourism. *Marine Policy, 82*, 104–113.
28. Asari, N., Suratman, M. N., Mohd Ayob, N. A., & Abdul Hamid, N. H. (2021). Mangrove as a natural barrier to environmental risks and coastal protection. In *Mangroves: Ecology, biodiversity and management* (pp. 305–322). Springer Singapore.

29. Hernández-Delgado, E. A. (2015). The emerging threats of climate change on tropical coastal ecosystem services, public health, local economies and livelihood sustainability of small islands: Cumulative impacts and synergies. *Marine Pollution Bulletin, 101*(1), 5–28.
30. Kiuru, P., D'Auria, M. V., Muller, C. D., Tammela, P., Vuorela, H., & Yli-Kauhaluoma, J. (2014). Exploring marine resources for bioactive compounds. *Planta Medica, 80*(14), 1234–1246.
31. Leary, D., Vierros, M., Hamon, G., Arico, S., & Monagle, C. (2009). Marine genetic resources: A review of scientific and commercial interest. *Marine Policy, 33*(2), 183–194.
32. Benitez-Nelson, C. R. (2000). The biogeochemical cycling of phosphorus in marine systems. *Earth-Science Reviews, 51*(1–4), 109–135.
33. Mwaipopo, R., & Shalli, M. (2016). *Aesthetic, cultural and spiritual services from coastal and marine environments*. United Nations.
34. Hernández, O. G. (2019). *New dialogues with the ocean as a recreational diver* (Doctoral dissertation, Aberystwyth University, Wales).
35. Vilkka, L. (1997). *The intrinsic value of nature* (Vol. 59). Rodopi.
36. Hagstrom, G. I., & Levin, S. A. (2017). Marine ecosystems as complex adaptive systems: Emergent patterns, critical transitions, and public goods. *Ecosystems, 20*, 458–476.
37. Bennett, N. J., Whitty, T. S., Finkbeiner, E., Pittman, J., Bassett, H., Gelcich, S., & Allison, E. H. (2018). Environmental stewardship: A conceptual review and analytical framework. *Environmental Management, 61*, 597–614.
38. Wang, B., Hua, L., Mei, H., Wu, X., Kang, Y., & Zhao, N. (2024). Impact of climate change on the dynamic processes of marine environment and feedback mechanisms: An overview. *Archives of Computational Methods in Engineering, 31*(6), 3377–3408.
39. Duterte, J. P. (2024). Global trends in marine biodiversity: Insights for conservation and sustainable management. *International Journal of Research and Innovation in Social Science, 8*(14), 474–480.
40. Hossain, M. M. (2019). Future importance of healthy oceans: Ecosystem functions and biodiversity, marine pollution, carbon sequestration, ecosystem goods and services. *Journal of Ocean and Coastal Economics, 6*(2), 4.
41. Crain, C. M., Halpern, B. S., Beck, M. W., & Kappel, C. V. (2009). Understanding and managing human threats to the coastal marine environment. *Annals of the New York Academy of Sciences, 1162*(1), 39–62.
42. Hughes, T. P., Bellwood, D. R., Folke, C., Steneck, R. S., & Wilson, J. (2005). New paradigms for supporting the resilience of marine ecosystems. *Trends in Ecology & Evolution, 20*(7), 380–386.
43. Valiela, I., & Valiela, I. (1995). *Marine ecological processes* (Vol. 686). Springer.
44. Forest, A., Stemmann, L., Picheral, M., Burdorf, L., Robert, D., Fortier, L., & Babin, M. (2012). Size distribution of particles and zooplankton across the shelf-basin system in Southeast Beaufort Sea: Combined results from an Underwater Vision Profiler and vertical net tows. *Biogeosciences, 9*(4), 1301–1320.
45. Stephenson, T. A., & Stephenson, A. (1949). The universal features of zonation between tide-marks on rocky coasts. *The Journal of Ecology, 37*, 289–305.
46. Leeuwis, R. H., & Gamperl, A. K. (2022). Adaptations and plastic phenotypic responses of marine animals to the environmental challenges of the high intertidal zone. In *Oceanography and marine biology* (pp. 625–679). CRC Press.
47. Archibold, O. W., & Archibold, O. W. (1995). Coastal and marine ecosystems. In *Ecology of world vegetation* (pp. 388–424). Springer Science & Business Media.
48. Beamish, R. J., McFarlane, G. A., & King, J. R. (2005). Migratory patterns of pelagic fishes and possible linkages between open ocean and coastal ecosystems off the Pacific coast of North America. *Deep Sea Research Part II: Topical Studies in Oceanography, 52*(5–6), 739–755.
49. Angel, M. V. (1997). 1 What is the deep sea? In *Fish physiology* (Vol. 16, pp. 1–41). Academic Press.

50. Hestetun, J. T., Sjøtun, K., Aksnes, D. L., Asplin, L., Devine, J., Falkenhaug, T., et al. (2017). The marine environment. In *Marine ecological field methods: A guide for marine biologists and fisheries scientists* (pp. 1–31). Wiley-Blackwell.
51. Hornnes, I. (2023). *The role of bioluminescent prey in the diet of two hatchetfish species and four lanternfish species in the North Atlantic Ocean in relation to eye-size* (Master's thesis, The University of Bergen).
52. Gowri, V., Thenmozhi, S., Raha, S., Selvakumar, P., & Khokale, S. K. (2025). Climate change and the deep sea. In *Technological advancements for deep sea ecosystem conservation and exploration* (pp. 73–98). IGI Global Scientific Publishing.
53. Waterman, T. H. (2001). Evolutionary challenges of extreme environments (part 2). *Journal of Experimental Zoology, 291*(2), 130–168.
54. Jamieson, A. (2015). *The hadal zone: Life in the deepest oceans*. Cambridge University Press.
55. Worm, B., & Lotze, H. K. (2021). Marine biodiversity and climate change. In *Climate change* (pp. 445–464). Elsevier.
56. Hoegh-Guldberg, O. (2011). Coral reef ecosystems and anthropogenic climate change. *Regional Environmental Change, 11*(Suppl 1), 215–227.
57. Arceo-Carranza, D., Chiappa-Carrara, X., Chávez López, R., & Yáñez Arenas, C. (2021). Mangroves as feeding and breeding grounds. In *Mangroves: Ecology, biodiversity and management* (pp. 63–95). Springer Singapore.
58. Whitfield, A. K. (2017). The role of seagrass meadows, mangrove forests, salt marshes and reed beds as nursery areas and food sources for fishes in estuaries. *Reviews in Fish Biology and Fisheries, 27*(1), 75–110.
59. Calbet, A. (2025). The future of marine plankton. In *The Ocean of Today, the Legacy of Tomorrow: Navigating the future of marine life and ecosystems* (pp. 77–85). Springer Nature Switzerland.
60. George, R. Y. (2020). Extremophilic life in our oceans as models for astrobiology. In *Extremophiles as astrobiological models* (pp. 161–182). Wiley.
61. Barnes, D. K., & Tarling, G. A. (2017). Polar oceans in a changing climate. *Current Biology, 27*(11), R454–R460.
62. Levin, L. A., Boesch, D. F., Covich, A., Dahm, C., Erséus, C., Ewel, K. C., et al. (2001). The function of marine critical transition zones and the importance of sediment biodiversity. *Ecosystems, 4*(5), 430–451.
63. McLean, R. F., Tsyban, A., Burkett, V., Codignotto, J. O., Forbes, D. L., Mimura, N., et al. (2001). Coastal zones and marine ecosystems. In *Climate change* (pp. 343–379). Cambridge University Press.
64. Thiagarajan, C., & Devarajan, Y. (2024). The urgent challenge of ocean pollution: Impacts on marine biodiversity and human health. *Regional Studies in Marine Science, 81*, 103995.
65. Nichols, C. R., Zinnert, J., & Young, D. R. (2018). Degradation of coastal ecosystems: Causes, impacts and mitigation efforts. In *Tomorrow's coasts: Complex and impermanent* (pp. 119–136). Springer International Publishing.
66. Laffoley, D., Baxter, J. M., Day, J. C., Wenzel, L., Bueno, P., & Zischka, K. (2019). Marine protected areas. In *World seas: An environmental evaluation* (pp. 549–569). Academic Press.
67. Knauss, J. A., & Garfield, N. (2016). *Introduction to physical oceanography*. Waveland Press.
68. Water, O. F. (2011). Physical properties of seawater.
69. Talley, L. D. (2002). Salinity patterns in the ocean. In *The Earth system: Physical and chemical dimensions of global environmental change* (Vol. 1, pp. 629–640). Wiley.
70. Rahmstorf, S. (2006). Thermohaline ocean circulation. In *Encyclopedia of quaternary sciences* (Vol. 5(18), p. 191). Elsevier.
71. Macdonald, A. G. (1997). Hydrostatic pressure as an environmental factor in life processes. *Comparative Biochemistry and Physiology Part A: Physiology, 116*(4), 291–297.
72. Tett, P. A. U. L. (1990). The photic zone. In *Light and life in the sea* (pp. 59–87). Cambridge University Press.

73. Kim, Y., & Lee, W. G. (2022). Seawater and its resources. In *Seawater batteries: Principles, materials and technology* (pp. 1–35). Springer Nature Singapore.
74. Mukherjee, S., Chanda, A., & Swapna/Ghosh Mukherjee (Kaushik Kira). (2023). Environmental oceanography and coastal dynamics.
75. Millero, F. J. (1974). The physical chemistry of seawater. *Annual Review of Earth and Planetary Sciences, 2*, 101.
76. Sakshaug, E., Rey, F., & Van Ardelan, M. (2009). Seawater, its constituents and chemistry. *Chap, 4*, 83–116.
77. Lewis, E. L., & Perkin, R. G. (1978). Salinity: Its definition and calculation. *Journal of Geophysical Research: Oceans, 83*(C1), 466–478.
78. Levasseur, M. E., & Therriault, J. C. (1987). Phytoplankton biomass and nutrient dynamics in a tidally induced upwelling. The role of the NO_3: SiO_4 ratio. *Marine Ecology Progress Series, 39*, 87–97.
79. Revelle, R., & Fairbridge, R. (1957). Carbonates and carbon dioxide.
80. Jiang, L. Q., Carter, B. R., Feely, R. A., Lauvset, S. K., & Olsen, A. (2019). Surface ocean pH and buffer capacity: Past, present and future. *Scientific Reports, 9*(1), 18624.
81. Van Gennip, S. J., Popova, E. E., Yool, A., Pecl, G. T., Hobday, A. J., & Sorte, C. J. (2017). Going with the flow: The role of ocean circulation in global marine ecosystems under a changing climate. *Global Change Biology, 23*(7), 2602–2617.
82. Huang, R. X. (2010). *Ocean circulation: Wind-driven and thermohaline processes*. Cambridge University Press.
83. Balasubramanian, A. (2014). The ocean currents. In *Proceedings of the Country-Wide Classroom Educational TV Programme-Gyan Darshan*, Mysore, India.
84. Rossby, T. (1999). On gyre interactions. *Deep Sea Research Part II: Topical Studies in Oceanography, 46*(1–2), 139–164.
85. Smith, K. L., Jr., Ruhl, H. A., Bett, B. J., Billett, D. S. M., Lampitt, R. S., & Kaufmann, R. S. (2009). Climate, carbon cycling, and deep-ocean ecosystems. *Proceedings of the National Academy of Sciences, 106*(46), 19211–19218.
86. Merino, M., & Monreal-Gómez, M. A. (2009). Ocean currents and their impact on marine life. *Marine Ecology, 1*, 47–52.
87. Anderson, T. R., & Lucas, M. I. (2009). Upwelling ecosystems. *Ecosystem Ecology, 450*.
88. Bryden, H. L., Robinson, C., & Griffiths, G. (2012). Changing currents: A strategy for understanding and predicting the changing ocean circulation. *Philosophical Transactions of the Royal Society A: Mathematical, Physical and Engineering Sciences, 370*(1980), 5461–5479.
89. Kämpf, J., & Chapman, P. (2016). *Upwelling systems of the world* (pp. 31–42). Springer International Publishing Switzerland.
90. Davarpanah Jazi, S., Wells, M. G., Peakall, J., Dorrell, R. M., Thomas, R. E., Keevil, G. M., et al. (2020). Influence of Coriolis force upon bottom boundary layers in a large-scale gravity current experiment: Implications for evolution of sinuous deep-water channel systems. *Journal of Geophysical Research: Oceans, 125*(3), e2019JC015284.
91. Chavez, F. P., & Messié, M. (2009). A comparison of eastern boundary upwelling ecosystems. *Progress in Oceanography, 83*(1–4), 80–96.
92. Kessler, W. S. (2006). The circulation of the eastern tropical Pacific: A review. *Progress in Oceanography, 69*(2–4), 181–217.
93. Capone, D. G., & Hutchins, D. A. (2013). Microbial biogeochemistry of coastal upwelling regimes in a changing ocean. *Nature Geoscience, 6*(9), 711–717.
94. Bakun, A., Black, B. A., Bograd, S. J., Garcia-Reyes, M., Miller, A. J., Rykaczewski, R. R., & Sydeman, W. J. (2015). Anticipated effects of climate change on coastal upwelling ecosystems. *Current Climate Change Reports, 1*(2), 85–93.
95. Toggweiler, J. R., & Key, R. M. (2001). Ocean circulation: Thermohaline circulation. *Encyclopedia of Atmospheric Sciences, 4*, 1549–1555.

96. Bigg, G. R., Jickells, T. D., Liss, P. S., & Osborn, T. J. (2003). The role of the oceans in climate. *International Journal of Climatology: A Journal of the Royal Meteorological Society, 23*(10), 1127–1159.
97. Harris, P. T., & Macmillan-Lawler, M. (2017). Origin and geomorphic characteristics of ocean basins. In *Submarine geomorphology* (pp. 111–134). Springer International Publishing.
98. Stanley, S. M. (2018). Ocean circulation: Conveyor of past and future climate. In *The Earth around us* (pp. 308–321). Routledge.
99. Hastenrath, S. (2012). *Climate and circulation of the tropics* (Vol. 8). Springer Science & Business Media.
100. Cindrella, J., & Renjith, K. R. (2025). Understanding the role of oceans in global sustainability: Biogeochemistry and challenges in the Anthropocene. In *Radiation status in the marine world* (pp. 1–29). Springer Nature Switzerland.
101. Zardi, D. (2024). Atmosphere and ocean interactions. *Rendiconti Lincei. Scienze Fisiche e Naturali, 35*(2), 311–325.
102. Wang, C., Deser, C., Yu, J. Y., DiNezio, P., & Clement, A. (2016). El Niño and southern oscillation (ENSO): A review. In *Coral reefs of the eastern tropical Pacific: Persistence and loss in a dynamic environment* (pp. 85–106). Springer Dordrecht.
103. Islam, F. S. (2025). Assessment of the Global Climatic Impacts due to El Nino and La Nina Events. *Journal of Global Ecology and Environment, 21*(3), 1–26.
104. Stine, A. R., & Huybers, P. (2012). Changes in the seasonal cycle of temperature and atmospheric circulation. *Journal of Climate, 25*(21), 7362–7380.
105. Ussiri, D. A., & Lal, R. (2017). *Carbon sequestration for climate change mitigation and adaptation* (pp. 287–325). Springer International Publishing.
106. Kennedy, V. S., Twilley, R. R., Kleypas, J. A., Cowan, J. H., Jr., & Hare, S. R. (2002). *Coastal and marine ecosystems & global climate change*. Pew Center on Global Climate Change.
107. Xie, S. P., Deser, C., Vecchi, G. A., Collins, M., Delworth, T. L., Hall, A., et al. (2015). Towards predictive understanding of regional climate change. *Nature Climate Change, 5*(10), 921–930.
108. National Academies of Sciences, Medicine, Division on Earth, Life Studies, Board on Atmospheric Sciences, Ocean Studies Board, & Committee on Sustaining Ocean Observations to Understand Future Changes in Earth's Climate. (2018). *Sustaining ocean observations to understand future changes in Earth's climate*. National Academies Press.
109. Liao, B., Xiao, B., & Li, Z. (2019). Coral reef ecosystem. In *Symbiotic microbiomes of coral reefs sponges and corals* (pp. 1–15). Springer Netherlands.
110. Martin-Garin, B., & Montaggioni, L. F. (2023). *Corals and reefs: From the beginning to an uncertain future*. Springer.
111. Glynn, P. W., & Enochs, I. C. (2010). Invertebrates and their roles in coral reef ecosystems. In *Coral reefs: An ecosystem in transition* (pp. 273–325). Springer Dordrecht.
112. Barbier, E. B., Hacker, S. D., Kennedy, C., Koch, E. W., Stier, A. C., & Silliman, B. R. (2011). The value of estuarine and coastal ecosystem services. *Ecological Monographs, 81*(2), 169–193.
113. Riegl, B., Bruckner, A., Coles, S. L., Renaud, P., & Dodge, R. E. (2009). Coral reefs: Threats and conservation in an era of global change. *Annals of the New York Academy of Sciences, 1162*(1), 136–186.
114. Lough, J. M., Anderson, K. D., & Hughes, T. P. (2018). Increasing thermal stress for tropical coral reefs: 1871–2017. *Scientific Reports, 8*(1), 1–8.
115. Ibrahim-Bathis, K. (2021). Insight to the spatial-temporal extent of mangrove forests in the northern coast of Kerala. In *Remote sensing of ocean and coastal environments* (pp. 375–392). Elsevier.
116. Ayyam, V., Palanivel, S., & Chandrakasan, S. (2019). Coastal ecosystems and services. In *Coastal ecosystems of the tropics—Adaptive management* (pp. 21–47). Springer Singapore.

117. Akram, H., Hussain, S., Mazumdar, P., Chua, K. O., Butt, T. E., & Harikrishna, J. A. (2023). Mangrove health: A review of functions, threats, and challenges associated with mangrove management practices. *Forests, 14*(9), 1698.
118. Vierros, M. (2017). Communities and blue carbon: The role of traditional management systems in providing benefits for carbon storage, biodiversity conservation and livelihoods. In *Climatic change* (Vol. 140(1), pp. 89–100). Springer.
119. Schubert, H., & Telesh, I. (2017). Estuaries and coastal lagoons. In *Biological oceanography of the Baltic Sea* (pp. 483–509). Springer Netherlands.
120. Cardoso, P. G. (2022). Estuaries: Dynamics, biodiversity, and impacts. In *Life below water* (pp. 355–366). Springer International Publishing.
121. Kennish, M. J. (2002). Environmental threats and environmental future of estuaries. *Environmental Conservation, 29*(1), 78–107.
122. Weinstein, M. P. (2008). Ecological restoration and estuarine management: Placing people in the coastal landscape. *Journal of Applied Ecology, 45*(1), 296–304.
123. Ramirez-Llodra, E., Brandt, A., Danovaro, R., De Mol, B., Escobar, E., German, C. R., et al. (2010). Deep, diverse and definitely different: Unique attributes of the world's largest ecosystem. *Biogeosciences, 7*(9), 2851–2899.
124. Fenolio, D. (2016). *Life in the dark: Illuminating biodiversity in the shadowy haunts of planet earth*. JHU Press.
125. Dick, G. J. (2019). The microbiomes of deep-sea hydrothermal vents: Distributed globally, shaped locally. *Nature Reviews Microbiology, 17*(5), 271–283.
126. Ramirez-Llodra, E., & Rogers, A. D. (2025). An ecosystem view of anthropogenic impacts in the deep ocean. In *Volume 2: Marine ecology* (pp. 201–228). CRC Press.
127. Glover, A. G., & Smith, C. R. (2003). The deep-sea floor ecosystem: Current status and prospects of anthropogenic change by the year 2025. *Environmental Conservation, 30*(3), 219–241.
128. Valls, A., Coll, M., & Christensen, V. (2015). Keystone species: Toward an operational concept for marine biodiversity conservation. *Ecological Monographs, 85*(1), 29–47.
129. Walsh, J. R., Carpenter, S. R., & Vander Zanden, M. J. (2016). Invasive species triggers a massive loss of ecosystem services through a trophic cascade. *Proceedings of the National Academy of Sciences, 113*(15), 4081–4085.
130. Srinivas, N., Venkatrayulu, C., Shanmugam, K. R., Suresh, S., Vijayakumar, P., & Chakrapani, I. S. (2024). Challenges and investigate the biodiversity and ecosystem: Role of keystone species in conservation strategies. *Bulletin of Pure & Applied Sciences-Zoology*.
131. Garibaldi, A., & Turner, N. (2004). Cultural keystone species: Implications for ecological conservation and restoration. *Ecology and Society, 9*(3), 1.
132. Moore, J. C., Berlow, E. L., Coleman, D. C., De Ruiter, P. C., Dong, Q., Hastings, A., et al. (2004). Detritus, trophic dynamics and biodiversity. *Ecology Letters, 7*(7), 584–600.
133. Andrews, D. L., Curutchet, C., & Scholes, G. D. (2011). Resonance energy transfer: Beyond the limits. *Laser & Photonics Reviews, 5*(1), 114–123.
134. Hallegraeff, G. M. (2010). Ocean climate change, phytoplankton community responses, and harmful algal blooms: A formidable predictive challenge. *Journal of Phycology, 46*(2), 220–235.
135. Buck, J. C., & Ripple, W. J. (2017). Infectious agents trigger trophic cascades. *Trends in Ecology & Evolution, 32*(9), 681–694.
136. Douglas, V. (2025). The impact of overfishing on shark populations and marine ecosystem.
137. Dorgham, M. M. (2013). Effects of eutrophication. In *Eutrophication: Causes, consequences and control: Volume 2* (pp. 29–44). Springer Netherlands.
138. Van der Putten, W. H., Macel, M., & Visser, M. E. (2010). Predicting species distribution and abundance responses to climate change: Why it is essential to include biotic interactions across trophic levels. *Philosophical Transactions of the Royal Society B: Biological Sciences, 365*(1549), 2025–2034.
139. Lapointe, N. W., Cooke, S. J., Imhof, J. G., Boisclair, D., Casselman, J. M., Curry, R. A., et al. (2014). Principles for ensuring healthy and productive freshwater ecosystems that support sustainable fisheries. *Environmental Reviews, 22*(2), 110–134.

140. Link, J. (2010). *Ecosystem-based fisheries management: Confronting tradeoffs*. Cambridge University Press.
141. Hrdina, A., & Romportl, D. (2017). Evaluating global biodiversity hotspots—Very rich and even more endangered. *Journal of Landscape Ecology, 10*(1), 108–115.
142. Myers, N. (1990). The biodiversity challenge: Expanded hot-spots analysis. *Environmentalist, 10*(4), 243–256.
143. Pott, R. (2016). Fundamentals and perspectives of geobotanical research in the twenty-first century. In *Vegetation structure and function at multiple spatial, temporal and conceptual scales* (pp. 529–548). Springer International Publishing.
144. Hoeksema, B. W. (2007). Delineation of the Indo-Malayan centre of maximum marine biodiversity: The Coral Triangle. In *Biogeography, time, and place: Distributions, barriers, and islands* (pp. 117–178). Springer Netherlands.
145. Veach, V., Moilanen, A., & Di Minin, E. (2017). Threats from urban expansion, agricultural transformation and forest loss on global conservation priority areas. *PLoS One, 12*(11), e0188397.
146. Peyre, G. (2021). Terrestrial biodiversity hotspots: Challenges and opportunities. *Life on Land*, 1–20.
147. Mittermeier, R. A., Turner, W. R., Larsen, F. W., Brooks, T. M., & Gascon, C. (2011). Global biodiversity conservation: The critical role of hotspots. In *Biodiversity hotspots* (pp. 3–22). Springer.
148. Lopoukhine, N., Crawhall, N., Dudley, N., Figgis, P., Karibuhoye, C., Laffoley, D., et al. (2012). Protected areas: Providing natural solutions to 21st century challenges. *SAPIENS Surveys and Perspectives Integrating Environment and Society, 5*(2), 116–131.
149. Balasubramanian, M., & Sangha, K. K. (2023). Valuing ecosystem services applying indigenous perspectives from a global biodiversity hotspot, the Western Ghats, India. *Frontiers in Ecology and Evolution, 11*, 1026793.
150. Jones, H. P., Nickel, B., Srebotnjak, T., Turner, W., Gonzalez-Roglich, M., Zavaleta, E., & Hole, D. G. (2020). Global hotspots for coastal ecosystem-based adaptation. *PLoS One, 15*(5), e0233005.
151. Kareiva, P., & Marvier, M. (2003). Conserving biodiversity coldspots: Recent calls to direct conservation funding to the world's biodiversity hotspots may be bad investment advice. *American Scientist, 91*(4), 344–351.

Chapter 2
Anthropogenic Impacts and Challenges

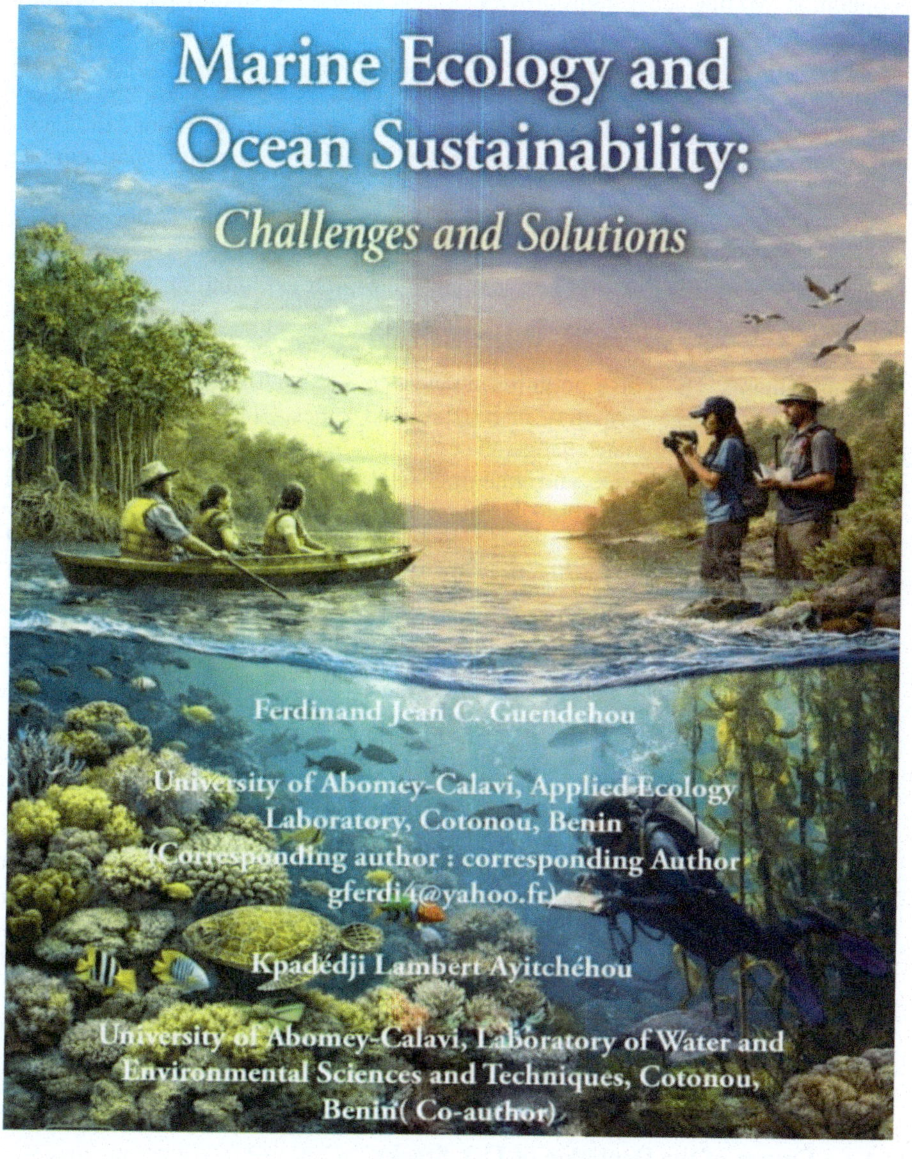

F. J. C. Guendehou, K. L. Ayitchéhou, *Marine Ecology and Ocean Sustainability*, https://doi.org/10.1007/978-3-032-32756-7_2

Understanding the scale and complexity of human impacts on marine ecosystems is essential for developing effective strategies to protect ocean health and ensure the sustainable use of marine resources. This research highlights the ecological, economic, and social consequences of anthropogenic activities in marine environments and underscores the urgent need for science-based management and international cooperation. Marine ecosystems provide critical services, including fishery production, climate regulation, coastal protection, and biodiversity conservation. However, the accelerating degradation of these systems threatens global food security, coastal economies, and the stability of marine ecological processes. By examining the key drivers of ocean degradation, this work contributes to the scientific foundation necessary for sustainable fisheries management, pollution mitigation, and climate adaptation policies.

This chapter examines the main anthropogenic impacts and environmental challenges that affect marine ecosystems. First, it analyzes the decline in fish stocks resulting from overfishing and unsustainable resource exploitation, including the impacts of bycatch and illegal fishing practices. Then it explores different forms of marine pollution, focusing on plastic pollution and microplastics, oil spills, chemical contaminants, and noise pollution, as well as the ecological consequences of eutrophication and the formation of coastal dead zones. Finally, this chapter discusses the impacts of climate change on ocean health, including ocean acidification, coral bleaching, sea level rise, and changing weather patterns. Together, these issues illustrate the complex interactions between human activities and marine systems, highlighting the urgent need for integrated management approaches and global cooperation to protect marine biodiversity and ensure long-term sustainability of the world's oceans.

1 Overfishing and Resource Exploitation

1.1 Decline of Fish Stocks

The decline in global fish stocks is a growing environmental, economic, and social concern [1]. Over the past century, the health of the world's oceans has been severely affected by unsustainable fishing practices, habitat destruction, pollution, and climate change [2]. Fish stocks defined as populations of fish species that are fished commercially have declined at alarming rates, posing serious threats to marine ecosystems and to millions of people who rely on seafood as a primary source of food and income [3].

1.1.1 Causes of the Depletion of Fish Stocks

The primary cause of the decline in fish stocks is overfishing [4]. Technological advancements in fishing gear and vessel capacity have made it possible to catch large amounts of fish with high efficiency [5]. Although this has supported the global demand for seafood, it has also pushed many species beyond their natural ability to replenish [6]. According to the Food and Agriculture Organization (FAO), more than 34% of global fish stocks are currently overfished, while nearly 60% are fished at their maximum sustainable levels [7].

1.1.2 Ecological and Economic Impacts

The ecological consequences of declining fish stocks are profound. When the top predator species are removed from the food chain, it disrupts marine ecosystems, leading to imbalances and the collapse of local biodiversity. For example, the decline of large predatory fish often results in population explosions of smaller fish and invertebrates, which can overgraze marine vegetation and destabilize the ecosystem [8]. Economically, the consequences are equally severe. Fisheries provide jobs for more than 200 million people worldwide, particularly in coastal and island nations [9]. As fish stocks decrease, catch sizes decrease, leading to loss of income, increased fish prices, and reduced access to affordable protein [10]. Small-scale and artisanal fishermen are particularly vulnerable, as they often lack the resources to travel further or switch to alternative livelihoods [11]. Food security is another critical concern. Fish is a vital source of animal protein for more than three billion people [12]. In many low-income countries, it accounts for more than 50% of dietary protein. The decline in fish availability can therefore exacerbate malnutrition and poverty, especially in regions already facing food insecurity [13].

1.1.3 Efforts to Reverse the Trend

Despite the challenges, there are strategies available to reverse the decline in fish stocks and restore marine ecosystems [14]. Sustainable fishing management is one of the most effective approaches. This includes implementing science-based catch limits, seasonal closures, gear restrictions, and quotas that ensure that fishing does not exceed the regenerative capacity of fish populations [15]. Marine protected areas (MPAs) offer another solution. These designated zones restrict or prohibit fishing, allowing ecosystems to recover and replenish surrounding areas. Studies have shown that MPAs, when effectively enforced, can increase biodiversity and fish biomass both within and outside protected boundaries [16]. International cooperation is essential for managing migratory species and transboundary stocks. Organizations such as Regional Fisheries Management Organizations (RFMOs) help coordinate sustainable practices across countries. However, its success depends on strong political will and compliance from the member nations [17]. Technological innovations also help to improve monitoring and enforcement. Satellite surveillance, electronic catch documentation, and traceability systems allow authorities to track fishing activity in real time, making it harder for illegal practices to go unnoticed [18]. Public awareness and market-based solutions can also support sustainable fishing. Certification programs like the Marine Stewardship Council (MSC) label encourage responsible consumer choices and incentivize producers to adopt sustainable methods [19].

In conclusion, the decline of global fish stocks is a critical issue that reflects the broader challenges of balancing human activity with ecological sustainability [20]. While the causes are complex, ranging from overfishing and pollution to climate change and governance failures, the solutions are within reach [21]. Restoring fish stocks is not only vital for ocean health but also for the well-being of communities that depend on them. With coordinated global action, informed management, and sustainable practices, the world's oceans can recover, ensuring a more secure future for both people and the planet [22].

1.2 Bycatch and Illegal Fishing

The sustainability of global fisheries is being increasingly threatened by two major challenges: bycatch and illegal, unreported, and unregulated (IUU) fishing. Both practices have far-reaching impacts on marine ecosystems, fish populations, economies, and food security [23].

Addressing them is crucial for ensuring healthy oceans and responsible management of marine resources.

1.2.1 Bycatch

Bycatch refers to the unintentional capture of non-target species during fishing operations. These can include juvenile fish, endangered species (such as sea turtles, sharks, and dolphins), seabirds, and even marine mammals [24]. Bycatch is a major issue in commercial fishing, particularly in methods, such as trawling, longlining, and gillnetting, which are often non-selective in what they catch. The scale of bycatch is staggering [25]. According to estimates by the Food and Agriculture Organization (FAO), bycatch may account for up to 40% of global marine catches, depending on the region and the fishing method [26]. Most of this unintended catch is discarded back into the ocean, often dead or dying, contributing to waste and ecological damage. Bycatch has several negative impacts, such as ecological damage, threat to endangered species, economic loss, and social and ethical concerns [27]. Reducing bycatch is possible with better management, technology, and regulation. Key strategies include selective fishing gear, bycatch limits and regulations, real-time monitoring, fishing area closures, and industry training and incentives [28].

1.2.2 Illegal, Unreported, and Unregulated (IUU) Fishing

IUU fishing refers to fishing activities that violate national or international laws or are not reported to relevant authorities [29]. It includes fishing without a license, fishing in closed areas or during off-seasons, catching protected species, falsifying catch documentation, and operating in another country's waters without permission [30]. Globally, IUU fishing accounts for an estimated 20–30% of all marine catches, with higher rates in some developing regions. It is valued at up to $23 billion annually, making it one of the largest illicit activities in the world [31]. IUU fishing undermines marine conservation efforts and has multiple damaging consequences, such as overexploitation of fish stocks, economic harm, threat to food security, and encouragement of crime networks [32].

Efforts to combat IUU fishing involve national and international cooperation, enhanced monitoring, and strong legal frameworks. Some effective approaches include:

1. Port State Measures: Under the FAO Port State Measures Agreement, countries can deny port access to vessels suspected of IUU fishing, making it harder for them to offload illegal catches [33].
2. Vessel Monitoring Systems (VMS): GPS tracking of fishing vessels allows authorities to detect unauthorized activity and monitor fishing routes in real time [34].
3. Blacklisting and Sanctions: Regional fisheries management organizations (RFMOs) can blacklist vessels or countries involved in IUU fishing and ban their seafood from international trade [35].
4. Catch Documentation Schemes: Requiring that catch data be recorded and verified ensures traceability and transparency throughout the seafood supply chain [36].
5. Consumer Awareness and Certification: Labels such as the Marine Stewardship Council (MSC) help consumers choose legally and sustainably sourced seafood, putting pressure on companies to act responsibly [37].

In general, bycatch and illegal fishing represent serious obstacles to the health and sustainability of global marine ecosystems. Together, they waste vast amounts of marine life, accelerate the depletion of fish stocks, and damage the livelihoods of legal fishermen [38]. Addressing these issues requires coordinated global action, robust enforcement, better fishing practices, and increased public awareness [39]. Through science-based management, technological innovation, and international cooperation, the tide can be turned toward a more sustainable and equitable use of the world's oceans [40].

1.3 Sustainable Fisheries Management

Sustainable fisheries management is a crucial strategy to ensure that marine and freshwater ecosystems remain productive, resilient, and capable of supporting biodiversity, food security, and livelihoods [41]. As global fish stocks face increasing threats from overfishing, habitat destruction, pollution, and climate change, the need for responsible, science-driven management has never been more urgent [42]. At its core, sustainable fisheries management involves the use of data-informed strategies and policies that balance ecological integrity with human needs. It aims to keep fish populations at healthy levels, minimize bycatch, protect marine habitats, and ensure that the exploitation of resources does not exceed the capacity to regenerate [43]. This requires not just limiting how much is caught, but also considering when, where, and how fishing takes place. Sustainable fisheries management is essential to protect ocean health while ensuring food, jobs, and economic stability [44]. Through science, innovation, inclusive governance, and international cooperation, the world's fisheries can recover and thrive [45]. With a shared commitment to sustainable practices, we can ensure that marine resources continue to benefit both people and the planet for generations to come [46].

2 Pollution in Marine Environments

2.1 Plastic Pollution and Microplastics

Plastic pollution is one of the most pressing environmental threats facing the world's oceans today. From vast floating debris fields to invisible microplastics, plastic waste has infiltrated nearly every corner of marine ecosystems. Its persistence, widespread distribution, and harmful effects on marine life and human health have made plastic pollution a global concern that requires urgent action [47]. Each year, 8–12 million metric tons of plastic are estimated to enter the oceans [48]. This pollution comes from a variety of sources, including poorly managed waste disposal, littering, fishing gear, industrial discharge, and stormwater runoff. Once in the ocean, plastic does not biodegrade, it breaks down into smaller fragments, leading to long-lasting contamination [49]. Plastic debris has been found in even the most remote marine environments, from Arctic ice to the deepest ocean trenches [50]. It is estimated that there are over 170 trillion plastic particles floating in the oceans of the world, and the numbers increase each year [51].

2.1.1 Macroplastics and Their Impacts

Macroplastics refer to larger plastic items, such as bottles, bags, packaging, fishing nets, and containers [52]. These materials pose immediate threats to marine wildlife:

- **Entanglement**: Sea turtles, dolphins, seals, and seabirds often get entangled in fishing lines, plastic bags, or six-pack rings, leading to injury, drowning, or death [53].
- **Ingestion**: Many marine animals mistake plastic for food. Turtles often eat plastic bags thinking they are jellyfish, while seabirds ingest floating plastic pieces, filling their stomachs with indigestible material that can cause starvation [54].
- **Habitat Damage**: Discarded fishing gear (ghost nets) continues to trap marine life long after being abandoned and can destroy coral reefs and seafloor habitats [38].

Beyond harming wildlife, floating plastic can transport invasive species to new environments, altering ecosystems, and threatening native biodiversity [55].

2.1.2 Microplastics: An Invisible Threat

Microplastics are plastic particles smaller than 5 mm [56]. They are either:

- **Primary Microplastics**: intentionally manufactured small (e.g., microbeads in cosmetics) [57].
- **Secondary Microplastics**: resulting from the breakdown of larger plastic debris through exposure to sunlight, waves, and abrasion [58].

Microplastics have become ubiquitous in all parts of the ocean, including surface waters, deep sea sediments, coastal areas, and even Arctic ice [59]. They are ingested by a wide range of marine organisms, from plankton to whales, often accumulating along the food chain [60].

Impacts of microplastics include:

- **Physical Harm**: Ingestion can block the digestive tract, reduce nutrient absorption, and lead to death in marine animals [61].
- **Chemical Exposure**: Microplastics can absorb and concentrate toxic pollutants such as PCBs, pesticides, and heavy metals from seawater. These toxins can leach into animal tissues and accumulate through trophic levels [62].
- **Human Health Concerns**: Humans are exposed to microplastics through seafood, drinking water, and even air. Although long-term health impacts are still being studied, the presence of plastic particles and associated chemicals in the human body raises serious concerns [63].

2.1.3 Solutions and Strategies

Addressing marine plastic pollution requires a comprehensive and coordinated response.

1. **Reduce Plastic Production and Use**
   The shift away from single-use plastics and the promotion of sustainable alternatives (such as paper, glass, or biodegradable materials) is critical [64].
2. **Improve Waste Management**
   Enhancing recycling, investing in waste collection infrastructure, and reducing landfill leakage can prevent plastic from reaching the oceans [65].
3. **Regulations and Bans**
   Many countries have implemented bans or restrictions on plastic bags, microbeads, and other single-use plastics. Strengthening these laws and enforcing international agreements such as the MARPOL Convention are essential [66].
4. **Clean-Up Initiatives**
   Projects such as Ocean Cleanup and coastal community beach cleanups in coastal communities help remove existing debris from the environment, although they must be paired with long-term prevention strategies to be effective [67].
5. **Innovation and Research**
   Developing biodegradable plastics, improving filtration systems to trap microplastics, and advancing materials science can reduce future pollution [68].

6. **Public Education and Behavior Change**
   Encouraging responsible consumer choices, such as reducing plastic use, properly recycling, and supporting sustainable products, plays a key role in tackling the problem [69].

In *conclusion*, plastic pollution and microplastics represent one of the most widespread and persistent threats to marine environments. They harm wildlife, disrupt ecosystems, and pose unknown risks to human health [70]. Combating this crisis requires global collaboration, stronger policies, scientific innovation, and public commitment. Only through a comprehensive change in how we produce, consume, and dispose of plastic can we protect the health of our oceans and our future.

2.2 *Oil Spills, Chemical Waste, and Noise Pollution*

Marine pollution, a major threat to oceanic ecosystems and coastal communities, arises from a variety of human activities [71]. Among the most critical and destructive forms are oil spills, chemical waste discharge, and noise pollution [72]. These not only disrupt marine biodiversity but also affect human livelihoods, food security, and climate regulation [73].

2.2.1 Oil Spills

Oil spills are among the most visible and devastating forms of marine pollution. They occur due to tanker accidents, pipeline ruptures, offshore drilling, and operational discharges [74]. Crude oil contains toxic hydrocarbons that are harmful to marine life. When spilled, oil spreads rapidly across the ocean surface, forming a slick that blocks sunlight, thus hindering photosynthesis in marine plants and phytoplankton [75]. Marine animals such as birds and mammals suffer from hypothermia when their fur or feathers become coated, losing their insulating ability. Ingestion of oil by fish and filter-feeding organisms leads to internal damage, reproductive failure, and death [76]. Long-term ecological impacts persist for decades in sediments and food chains. Major incidents like the 2010 Deepwater Horizon spill have demonstrated the scale of destruction, affecting thousands of species and altering ecosystems in vast regions [77].

2.2.2 Chemical Waste

Industrial, agricultural, and urban sources discharge numerous chemicals into the ocean, either directly or through rivers and runoff [49]. These include heavy metals (such as mercury, lead, and cadmium), persistent organic pollutants (such as PCBs and DDT), pharmaceuticals, pesticides, and nutrients, such as nitrogen and phosphorus. Once in the marine environment, these substances can be highly toxic even at low concentrations [78]. Heavy metals bioaccumulate in marine organisms and biomagnify up the food chain, ultimately threatening predators and humans [79]. Nutrient overload leads to eutrophication, promoting harmful algal blooms that produce toxins and create hypoxic or "dead" zones where oxygen levels are too low to support most marine life [80]. Coral reefs and seagrasses are especially vulnerable, suffering bleaching and degradation. The growing problem of plastic pollution also overlaps with chemical pollution, as plastics leach harmful additives and act as vectors for other toxic compounds [62].

2.2.3 Noise Pollution

Marine noise pollution, though less visible, significantly disrupts the natural behavior and communication of ocean life. The main sources include commercial shipping, naval sonar, oil and gas exploration (especially seismic surveys), and construction activities, such as pile driving [81]. Marine mammals such as whales and dolphins rely on acoustic and echolocation signals for navigation, feeding, and social interaction. Excess noise interferes with these processes, causing disorientation, behavioral changes, habitat displacement, and even physical injury, such as hearing loss or strandings [82]. Chronic noise can raise stress hormone levels in marine species, affecting immunity and reproduction. Invertebrates and fish are also affected, with altered spawning and migratory patterns [83]. As global maritime traffic increases, the level of ambient noise in the ocean has increased significantly, challenging the acoustic integrity of marine habitats [84].

2.2.4 Integrated Impact and Response

These types of pollution often interact synergistically. For example, an oil spill can release both physical and chemical contaminants while causing intense noise from cleanup operations [85]. Collectively, they weaken the resilience and threaten biodiversity. Addressing marine pollution requires international cooperation and strict enforcement of environmental regulations, such as MARPOL (for shipborne pollution), the London Convention (for ocean dumping), and regional agreements [86]. Technologies such as satellite monitoring, bioremediation, green chemistry, and quieter ship designs offer hope. Public awareness and corporate accountability are also essential to reduce pollution at the source and protect the health of our oceans [87].

2.3 *Eutrophication and Dead Zones*

Eutrophication is a form of marine pollution that arises from the excessive enrichment of water bodies with nutrients, primarily nitrogen and phosphorus. These nutrients typically originate from agricultural runoff, sewage discharge, industrial effluents, and atmospheric deposition [88]. Although nutrients are essential for aquatic life, their overabundance disrupts the natural balance of marine ecosystems, leading to a cascade of negative effects that culminate in the formation of hypoxic areas known as "dead zones" [89].

2.3.1 Causes and Mechanisms

The main sources of nutrient pollution are synthetic fertilizers used in agriculture, untreated or partially treated wastewater, livestock manure, and fossil fuel combustion [90]. When these nutrients enter rivers and coastal waters, they stimulate the rapid growth of phytoplankton and algae. This process, known as algal bloom, may initially increase productivity but quickly turns harmful [91]. As dense algal populations die and sink, their decomposition by bacteria consumes a large amount of dissolved oxygen in the water [92]. This depletion of oxygen, particularly in stratified or poorly mixed waters, creates hypoxic (low oxygen) or anoxic (no oxygen) conditions unsuitable for most marine organisms [89].

2.3.2 Dead Zones

Dead zones are areas in oceans or large lakes where oxygen levels fall below the threshold necessary to support most marine life, usually below 2 mg/L [93]. Organisms that cannot flee, such as shellfish or bottom-dwelling species, die in massive numbers. Mobile species such as fish may escape, but their migration disrupts food chains and fisheries [93]. These hypoxic zones can persist for weeks, months, or even become seasonal phenomena. The largest dead zone in the United States is formed annually in the Gulf of Mexico, fueled by nutrient-laden runoff from the Mississippi River basin [94]. Globally, more than 500 dead zones have been documented, affecting an area of more than 245,000 square kilometers, many more likely undetected [95].

2.3.3 Ecological and Economic Impacts

Eutrophication and dead zones severely alter the structure and function. Oxygen depletion kills benthic organisms, reduces biodiversity, and changes the compositions of the community to more tolerant, often less desirable species, such as jellyfish [96]. Decomposing algal matter releases additional toxins and greenhouse gases, such as methane and nitrous oxide, contributing to climate change. Coral reefs and seagrass beds are particularly sensitive to nutrient stress, suffering bleaching, smothering, and eventual collapse [97]. Economically, eutrophication undermines commercial and recreational fisheries, aquaculture, and tourism. Clean-up costs, fisheries losses, and healthcare expenses related to toxic algal blooms can be billions annually [98].

2.3.4 Harmful Algal Blooms (HABs)

A notable consequence of eutrophication is the proliferation of harmful algae bloom species of algae that produce potent toxins. These toxins can contaminate seafood, cause mass death of marine fauna, and pose serious health risks to humans, including neurological and respiratory illnesses [98]. HABs also reduce water clarity, clog fish gills, and create scums and foul odors, making water bodies unusable for recreation [99].

2.3.5 Mitigation and Management

Combatting eutrophication and dead zones requires a multi-pronged approach. Key strategies include:

- **Nutrient Management**: Reducing fertilizer application, promoting precision agriculture, and improving livestock waste management to limit runoff [100].
- **Wastewater Treatment**: Upgrading sewage treatment plants to remove nutrients more effectively before discharge [101].
- **Restoration**: Reestablishing wetlands and vegetative buffer zones to filter runoff and absorb excess nutrients [102].
- **Policy and Regulation**: Enforcement of international and national regulations such as the Clean Water Act (USA), HELCOM agreements (Baltic Sea), and the EU Nitrates Directive [103].
- **Monitoring and Forecasting**: Remote sensing, modeling, and real-time monitoring to detect early signs of eutrophication and manage bloom events [104].
- **Public Awareness**: Educating communities and stakeholders on sustainable land use and responsible chemical use [105].

Eutrophication and dead zones represent a clear example of how terrestrial activities can severely damage marine environments [106]. Their widespread occurrence, exacerbated by climate change and continued nutrient loading, threatens ocean health, food security, and coastal economies [107]. The solution to this issue depends on integrated watershed management, international collaboration, and a strong commitment to sustainable practices.

3 Climate Change and Ocean Health

3.1 Ocean Acidification

Ocean acidification is a major consequence of climate change and poses a significant threat to marine ecosystems. As atmospheric CO_2 levels increase due to human activities such as fossil fuel burning and deforestation, approximately 25–30% of this excess CO_2 is absorbed by the oceans [108]. When CO_2 dissolves in seawater, it forms carbonic acid, which lowers the pH and reduces carbonate ion key components for shell and skeleton formation in marine organisms [109]. Since the Industrial Revolution, the pH of the ocean surface has dropped by about 0.1 units, representing a 30% increase in acidity [110]. If emissions continue, the pH could drop by up to 0.4 units by 2100. This chemical change disrupts calcifying organisms, such as corals, mollusks, and some plankton, making it harder for them to build and maintain shells and skeletons [111]. Coral reefs, already stressed by warming, face compounded risks as acidification weakens reef structures and reduces resilience [112]. Beyond calcifiers, ocean acidification can affect fish behavior, sensory abilities, and reproduction. The entire food webs can destabilize as foundational species decline [113]. Additionally, acidification interacts with other stressors, warming, deoxygenation, and pollution, amplifying impacts across ecosystems. Mitigating ocean acidification requires urgent reductions in CO_2 emissions globally [114]. Marine protected areas, ecosystem-based management, and research on adaptive species and restoration techniques can also help buffer their effects. Ultimately, addressing ocean acidification is inseparable from tackling climate change itself [115].

4 Coral Bleaching

Coral bleaching is a stark indicator of the impact of climate change on ocean health. It occurs when corals, stressed primarily by elevated sea temperatures, expel symbiotic algae (zooxanthellae) that live in their tissues [116]. These algae provide corals with food through photosynthesis and give them their vibrant color. Without them, corals turn white "bleached" and are left weakened, more vulnerable to disease, and at risk of death if stressful conditions persist [117]. The main driver of mass bleaching events is ocean warming. Even a sustained rise of just 1–2 °C above average can trigger bleaching [118]. Climate change has increased the frequency, intensity, and duration of marine heatwaves, making recovery between events increasingly difficult. Global bleaching events were recorded in 1998, 2010, and 2014–2017, affecting reefs around the world, including the Great Barrier Reef, where repeated bleaching has caused widespread coral mortality [119]. Other stressors, such as ocean acidification, pollution, and overfishing, compound the effects of warming, reducing coral resilience. As corals die, reef ecosystems home to about 25% of all marine species begin to collapse, threatening biodiversity, fishing, and coastal protection [120]. Solutions include aggressive global action to reduce greenhouse gas emissions, protecting and restoring reef habitats, controlling local stressors, and supporting coral adaptation efforts, such as selective breeding and assisted gene flow [121]. However, without rapid climate mitigation, the future of coral reefs remains in critical jeopardy [122].

4.1 Rise and Changing Weather Patterns

Climate change is reshaping ocean health through rising sea levels and altered weather patterns, with far-reaching ecological, economic, and social impacts [123]. Driven primarily by thermal expansion of seawater and the melting of glaciers and ice sheets, global sea levels have increased by more than 20 cm since 1900, with the rate accelerating in recent decades [124]. Projections suggest a further rise of 0.3–1 m or more by 2100, depending on emission scenarios [125].

4.1.1 Sea-Level Rise

Rising seas threaten coastal and marine ecosystems, submerging wetlands, mangroves, and coral reefs that serve as vital buffers against storms and erosion [126]. Saltwater intrusion into freshwater systems disrupts estuaries, harms agriculture, and degrades drinking water supplies [127]. For marine life, altered salinity and habitat loss can lead to loss of biodiversity and migration of species to more favorable environments [128]. Furthermore, low-lying nations and island communities face displacement, infrastructure damage, and loss of livelihoods, particularly in fishing and tourism [129].

4.1.2 Changing Weather Patterns

Climate change intensifies weather variability, altering ocean currents, precipitation patterns, and storm behavior. Warmer ocean temperatures fuel stronger tropical cyclones and hurricanes, leading to increased coastal flooding and habitat destruction [130]. Changes in monsoons, wind patterns, and ocean-atmosphere interactions such as El Niño and La Niña disrupt marine ecosystems by changing nutrient flows, upwelling zones, and fish migration routes [131]. These disruptions can destroy fisheries, strain food security, and increase competition for marine resources. Together, the rise and shifting weather patterns weaken the resilience of marine systems already stressed by overfishing, pollution, and acidification [132]. Addressing these changes requires global emissions reductions, adaptive coastal planning, restoration of natural buffers (such as mangroves and reefs), and improved early warning systems for climate-driven disasters. Long-term ocean health depends on both mitigation and adaptation to these evolving climate threats [133] (Figs. 2.1, 2.2, 2.3, 2.4 and 2.5).

Fig. 2.1 Coral reef restoration through coral transplantation. Scientists transplant coral fragments to degraded reefs to enhance biodiversity recovery. **Source image**: NOAA Coral Reef Conservation Program. **Reference**: Edwards, A.J. (2010). Reef Rehabilitation Manual. Coral Reef Targeted Research Program

Fig. 2.2 Kelp forest ecosystems supporting marine biodiversity. Kelp forests provide essential habitat and contribute to coastal productivity. **Source image**: National Geographic Image Collection. **Reference**: Steneck et al. (2002). Kelp forest ecosystems: biodiversity and resilience. Environmental conservation

(a)

(b)

Fig. 2.3 Marine scientists monitoring reef restoration sites. Scientific monitoring ensures the success of restoration projects. (**a**) Mid-water coral nursery system designed to enhance fragment survival and growth under controlled environmental conditions before outplanting. (**b**) Field-based coral transplantation and ecological monitoring, including data collection on colony health, survival rates, and structural development of restored reef assemblages. **Source image**: NOAA Ocean Exploration. **Reference**: Duarte et al. (2020). Rebuilding marine life. Nature

Fig. 2.4 Seagrass meadows as critical marine carbon sinks. Seagrasses contribute significantly to carbon sequestration and marine biodiversity. **Source image**: Blue-Carbon Initiative. **Reference**: McLeod, E. et al. (2011). A blueprint for blue carbon. Frontiers in Ecology and the Environment

Fig. 2.5 Mangrove root systems that support marine biodiversity. Complex root structures provide nursery habitats for fish and crustaceans. **Source image**: Conservation International Image Library. **Reference**: Alongi, D.M. (2014). Carbon cycling and storage in mangrove forests. Annual Review of Marine Science

References

1. Clausen, R., & York, R. (2008). Global biodiversity decline of marine and freshwater fish: A cross-national analysis of economic, demographic, and ecological influences. *Social Science Research, 37*(4), 1310–1320.
2. Nellemann, C., & Hain, S. (2008). *In dead water: Merging of climate change with pollution, over-harvest, and infestations in the world's fishing grounds*. UNEP/Earthprint.
3. Arthington, A. H., Dulvy, N. K., Gladstone, W., & Winfield, I. J. (2016). Fish conservation in freshwater and marine realms: Status, threats and management. *Aquatic Conservation: Marine and Freshwater Ecosystems, 26*(5), 838–857.
4. Rose, G. A. (2004). Reconciling overfishing and climate change with stock dynamics of Atlantic cod (Gadus morhua) over 500 years. *Canadian Journal of Fisheries and Aquatic Sciences, 61*(9), 1553–1557.
5. Eigaard, O. R., Marchal, P., Gislason, H., & Rijnsdorp, A. D. (2014). Technological development and fisheries management. *Reviews in Fisheries Science & Aquaculture, 22*(2), 156–174.
6. Belton, B., Reardon, T., & Zilberman, D. (2020). Sustainable commoditization of seafood. *Nature Sustainability, 3*(9), 677–684.
7. Melnychuk, M. C., Baker, N., Hively, D., Mistry, K., Pons, M., Ashbrook, C. E., et al. (2020). *Global trends in status and management of assessed stocks: Achieving sustainable fisheries through effective management* (Vol. 665). Food & Agriculture Organization.
8. Marques, L. (2020). Collapse of biodiversity in the aquatic environment. In *Capitalism and environmental collapse* (pp. 275–301). Springer International Publishing.
9. Teh, L. C., & Sumaila, U. R. (2013). Contribution of marine fisheries to worldwide employment. *Fish and Fisheries, 14*(1), 77–88.
10. Delgado, C. L., Wada, N., Rosegrant, M. W., Meijer, S., & Ahmed, M. (2003). *Outlook for fish to 2020: Meeting global demand* (Vol. 15). International Food Policy Research Institute.
11. March, A., & Failler, P. (2022). Small-scale fisheries development in Africa: Lessons learned and best practices for enhancing food security and livelihoods. *Marine Policy, 136*, 104925.
12. Tidwell, J. H., & Allan, G. L. (2001). Fish as food: Aquaculture's contribution. *The EMBO Reports, 2*, 958.
13. Bennett, A., Patil, P., Kleisner, K., Rader, D., Virdin, J., & Basurto, X. (2018). Contribution of fisheries to food and nutrition security: Current knowledge, policy, and research. *NI Report, 18*, 2.
14. Murawski, S. A. (2010). Rebuilding depleted fish stocks: The good, the bad, and, mostly, the ugly. *ICES Journal of Marine Science, 67*(9), 1830–1840.
15. Petza, D., & Katsanevakis, S. (2024). Science-informed recommendations to enhance the effectiveness of area-based fisheries management for fisheries sustainability and marine conservation: A global mini-review. *Fisheries Research, 272*, 106947.
16. Weigel, J. Y., Mannle, K. O., Bennett, N. J., Carter, E., Westlund, L., Burgener, V., et al. (2014). Marine protected areas and fisheries: Bridging the divide. *Aquatic Conservation: Marine and Freshwater Ecosystems, 24*(S2), 199–215.
17. Løbach, T., Petersson, M., Haberkon, E., & Mannini, P. (2020). Regional fisheries management organizations and advisory bodies: Activities and developments, 2000–2017. *FAO Fisheries and Aquaculture Technical Paper*, (651), I-112.
18. Probst, W. N. (2020). How emerging data technologies can increase trust and transparency in fisheries. *ICES Journal of Marine Science, 77*(4), 1286–1294.
19. Ponte, S. (2012). The Marine Stewardship Council (MSC) and the making of a market for 'sustainable fish'. *Journal of Agrarian Change, 12*(2–3), 300–315.
20. Blanchard, J. L., Watson, R. A., Fulton, E. A., Cottrell, R. S., Nash, K. L., Bryndum-Buchholz, A., et al. (2017). Linked sustainability challenges and trade-offs among fisheries, aquaculture and agriculture. *Nature Ecology & Evolution, 1*(9), 1240–1249.

21. Khan, A. S., & Neis, B. (2010). The rebuilding imperative in fisheries: Clumsy solutions for a wicked problem? *Progress in Oceanography, 87*(1–4), 347–356.
22. Fleming, L. E., Landrigan, P. J., Ashford, O. S., Whitman, E. M., Swift, A., Gerwick, W. H., et al. (2024). Enhancing human health and wellbeing through sustainably and equitably unlocking a healthy ocean's potential. *Annals of Global Health, 90*(1), 41.
23. Rosello, M. (2020). Illegal, unreported and unregulated (IUU) fishing as a maritime security concern. In *Global challenges in maritime security: An introduction* (pp. 33–47). Springer International Publishing.
24. Sacchi, J. (2021). *Overview of mitigation measures to reduce the incidental catch of vulnerable species in fisheries*. FAO.
25. Favaro, B. (2013). Can fishing gear protect non-target fish? Design and evaluation of bycatch reduction technology for commercial fisheries.
26. Clarke, S., Sato, M., Small, C., Sullivan, B., Inoue, Y., & Ochi, D. (2014). Bycatch in longline fisheries for tuna and tuna-like species: A global review of status and mitigation measures. *FAO Fisheries and Aquaculture Technical Paper, 588*, 1–199.
27. Gray, C. A., & Kennelly, S. J. (2018). Bycatches of endangered, threatened and protected species in marine fisheries. *Reviews in Fish Biology and Fisheries, 28*(3), 521–541.
28. Gilman, E. L. (2011). Bycatch governance and best practice mitigation technology in global tuna fisheries. *Marine Policy, 35*(5), 590–609.
29. Young, M. A. (2016). International trade law compatibility of market-related measures to combat illegal, unreported and unregulated (IUU) fishing. *Marine Policy, 69*, 209–219.
30. Lemrabott, S., van Leeuwen, A., Piersma, T., Braham, C. B., Ball, A. C., Araujo, A., et al. (2024). The chronology of overfishing in a remote West-African coastal ecosystem. *Ecology and Society, 29*(1), 9.
31. Metuzals, K., Baird, R., Pitcher, T., Sumaila, U. R., & Ganapathiraju, P. (2010). One fish, two fish, IUU, and no fish: Unreported fishing worldwide. In *Handbook of marine fisheries conservation and management* (pp. 165–181). Oxford University Press.
32. Cisneros-Montemayor, A. M., Cisneros-Mata, M. A., Harper, S., & Pauly, D. (2013). Extent and implications of IUU catch in Mexico's marine fisheries. *Marine Policy, 39*, 283–288.
33. Molenaar, E. J. (2010). Port State Jurisdiction to combat IUU fishing: The Port State measures agreement.
34. Enguehard, R. A., Devillers, R., & Hoeber, O. (2013). Comparing interactive and automated mapping systems for supporting fisheries enforcement activities—A case study on vessel monitoring systems (VMS). *Journal of Coastal Conservation, 17*(1), 105–119.
35. Scanlon, Z. (2019). Safeguarding the legitimacy of illegal, unreported and unregulated fishing vessel listings. *International & Comparative Law Quarterly, 68*(2), 369–398.
36. Hopkins, C. R., Roberts, S. I., Caveen, A. J., Graham, C., & Burns, N. M. (2024). Improved traceability in seafood supply chains is achievable by minimising vulnerable nodes in processing and distribution networks. *Marine Policy, 159*, 105910.
37. Jafarov, N. (2023). The role of the marine stewardship council label on consumers' purchasing behaviour.
38. Carneiro, M., & Martins, R. (2021). Destructive fishing practices and their impact on the marine ecosystem. In *Life below water* (pp. 1–11). Springer International Publishing.
39. Österblom, H., & Bodin, Ö. (2012). Global cooperation among diverse organizations to reduce illegal fishing in the Southern Ocean. *Conservation Biology, 26*(4), 638–648.
40. Caldeira, M., Sekinairai, A. T., & Vierros, M. (2025). Weaving science and traditional knowledge: Toward sustainable solutions for ocean management. *Marine Policy, 174*, 106591.
41. Lapointe, N. W., Cooke, S. J., Imhof, J. G., Boisclair, D., Casselman, J. M., Curry, R. A., et al. (2014). Principles for ensuring healthy and productive freshwater ecosystems that support sustainable fisheries. *Environmental Reviews, 22*(2), 110–134.
42. Abeysinghe, K. C., Weerarathne, A. G. P. T., Colonne, D. K. C., & Bellanthudawa, B. K. A. (2025). Sustainable development goals (SDGs), climate change, and the development of aquaculture and fisheries industries. *Environmental Reviews, 33*, 1.

43. Cafiero, C., Blanco Fonseca, M., Chelle, M., Egamberdieva, D., Kanter, R., Kimani, S., ... & Shyam, S. (2022). Data collection and analysis tools for food security and nutrition. Towards enhancing effective, inclusive, evidence-informed, decision making.
44. Sumaila, U. R., Bellmann, C., & Tipping, A. (2016). Fishing for the future: An overview of challenges and opportunities. *Marine Policy, 69*, 173–180.
45. Darias, M. J., Brink-Hull, M., Bach, P., & Macey, B. M. (2024). Accelerating transformation in aquatic food systems through holistic and inclusive innovation: Building and strengthening innovation ecosystems.
46. Fleming, L. E., Maycock, B., White, M. P., & Depledge, M. H. (2019). Fostering human health through ocean sustainability in the 21st century. *People and Nature, 1*(3), 276–283.
47. Islam, F. S. (2025). The effects of plastic and microplastic waste on the marine environment and the ocean. *European Journal of Environment and Earth Sciences, 6*(3), 1–9.
48. Rhodes, C. J. (2019). Solving the plastic problem: From cradle to grave, to reincarnation. *Science Progress, 102*(3), 218–248.
49. Gunasekara, M. I., Mahawaththa, I., Madhubhashini, D., & Amarasena, K. (2025). Pollution from land-based sources: Industrial and urban runoff. In *Coastal and marine pollution: Source to sink, mitigation and management* (pp. 27–44). Wiley.
50. Bergmann, M., Collard, F., Fabres, J., Gabrielsen, G. W., Provencher, J. F., Rochman, C. M., et al. (2022). Plastic pollution in the Arctic. *Nature Reviews Earth & Environment, 3*(5), 323–337.
51. Eriksen, M., Cowger, W., Erdle, L. M., Coffin, S., Villarrubia-Gómez, P., Moore, C. J., et al. (2023). A growing plastic smog, now estimated to be over 170 trillion plastic particles afloat in the world's oceans—Urgent solutions required. *PLoS One, 18*(3), e0281596.
52. Lechthaler, S., Waldschläger, K., Stauch, G., & Schüttrumpf, H. (2020). The way of macroplastic through the environment. *Environments, 7*(10), 73.
53. Hogan, E., & Warlick, A. (2017). Packing bands entangling pinnipeds around the world: Global review and policy. *Journal of International Wildlife Law & Policy, 20*(1), 96–104.
54. Ryan, P. G. (2016). Ingestion of plastics by marine organisms. In *Hazardous chemicals associated with plastics in the marine environment* (pp. 235–266). Springer International Publishing.
55. García-Gómez, J. C., Garrigós, M., & Garrigós, J. (2021). Plastic as a vector of dispersion for marine species with invasive potential. A review. *Frontiers in Ecology and Evolution, 9*, 629756.
56. Lim, X. (2021). Microplastics are everywhere—But are they harmful. *Nature, 593*(7857), 22–25.
57. Giustra, M., Sinesi, G., Spena, F., De Santes, B., Morelli, L., Barbieri, L., et al. (2024). Microplastics in cosmetics: Open questions and sustainable opportunities. *ChemSusChem, 17*(22), e202401065.
58. Corcoran, P. L. (2022). Degradation of microplastics in the environment. In *Handbook of microplastics in the environment* (pp. 531–542). Springer International Publishing.
59. Lusher, A. L., Tirelli, V., O'Connor, I., & Officer, R. (2015). Microplastics in Arctic polar waters: The first reported values of particles in surface and sub-surface samples. *Scientific Reports, 5*(1), 14947.
60. Roman, J., Estes, J. A., Morissette, L., Smith, C., Costa, D., McCarthy, J., et al. (2014). Whales as marine ecosystem engineers. *Frontiers in Ecology and the Environment, 12*(7), 377–385.
61. Negahdari, S., Sabaghan, M., Pirhadi, M., Alikord, M., Sadighara, P., Darvishi, M., & Nazer, M. (2021). Potential harmful effects of heavy metals as a toxic and carcinogenic agent in marine food—An overview. *Egyptian Journal of Veterinary Sciences, 52*(3), 379–385.
62. Kaur, S., Thakur, H., Singh, A., Ramasamy, V., & Mudgal, G. (2024). Poisoned seas: Chemical threats to marine life and human health. In *Sustainable development goals towards environmental toxicity and green chemistry: Environment and sustainability* (pp. 167–200). Springer Nature Switzerland.

63. Kannan, K., & Vimalkumar, K. (2021). A review of human exposure to microplastics and insights into microplastics as obesogens. *Frontiers in Endocrinology, 12*, 724989.
64. Viera, J. S., Marques, M. R., Nazareth, M. C., Jimenez, P. C., & Castro, Í. B. (2020). On replacing single-use plastic with so-called biodegradable ones: The case with straws. *Environmental Science & Policy, 106*, 177–181.
65. Kosior, E., & Crescenzi, I. (2020). Solutions to the plastic waste problem on land and in the oceans. In *Plastic waste and recycling* (pp. 415–446). Academic Press.
66. Bello, F. A. (2022). Critical analysis of policies on single-use plastics disposal from ships as a source of pollution to the marine environment.
67. Battisti, C., Poeta, G., Romiti, F., & Picciolo, L. (2020). Small environmental actions need of problem-solving approach: Applying project management tools to beach litter clean-ups. *Environments, 7*(10), 87.
68. Gao, W., Zhang, Y., Mo, A., Jiang, J., Liang, Y., Cao, X., & He, D. (2022). Removal of microplastics in water: Technology progress and green strategies. *Green Analytical Chemistry, 3*, 100042.
69. Aithal, P. S. (2024). Strategic and sustainable plastic management: Promoting awareness over bans for responsible usage. *Poornaprajna International Journal of Philosophy & Languages (PIJPL), 1*(1), 136–170.
70. Ghosh, S., Sinha, J. K., Ghosh, S., Vashisth, K., Han, S., & Bhaskar, R. (2023). Microplastics as an emerging threat to the global environment and human health. *Sustainability, 15*(14), 10821.
71. Crain, C. M., Halpern, B. S., Beck, M. W., & Kappel, C. V. (2009). Understanding and managing human threats to the coastal marine environment. *Annals of the New York Academy of Sciences, 1162*(1), 39–62.
72. Ogolo, N. A., Anih, O. C., & Onyekonwu, M. O. (2022). Sources and effects of environmental pollution from oil and gas industrial operations. *Arabian Journal of Chemical and Environmental Research, 9*(1), 98–121.
73. Rice, J. C., & Garcia, S. M. (2011). Fisheries, food security, climate change, and biodiversity: Characteristics of the sector and perspectives on emerging issues. *ICES Journal of Marine Science, 68*(6), 1343–1353.
74. Zhang, B., Matchinski, E. J., Chen, B., Ye, X., Jing, L., & Lee, K. (2019). Marine oil spills—Oil pollution, sources and effects. In *World seas: An environmental evaluation* (pp. 391–406). Academic Press.
75. Saadoun, I. M. (2015). Impact of oil spills on marine life. In *Emerging pollutants in the environment—Current and further implications* (Vol. 10, pp. 60–455). IntechOpen.
76. Thakur, A., & Koul, B. (2022). Impact of oil exploration and spillage on marine environments. In *Advances in oil-water separation* (pp. 115–135). Elsevier.
77. Fisher, C. R., Montagna, P. A., & Sutton, T. T. (2016). How did the Deepwater Horizon oil spill impact deep-sea ecosystems? *Oceanography, 29*(3), 182–195.
78. Weldeslassie, T., Naz, H., Singh, B., & Oves, M. (2017). Chemical contaminants for soil, air and aquatic ecosystem. In *Modern age environmental problems and their remediation* (pp. 1–22). Springer International Publishing.
79. Oros, A. (2025). Bioaccumulation and trophic transfer of heavy metals in marine fish: Ecological and ecosystem-level impacts. *Journal of Xenobiotics, 15*(2), 59.
80. Wurtsbaugh, W. A., Paerl, H. W., & Dodds, W. K. (2019). Nutrients, eutrophication and harmful algal blooms along the freshwater to marine continuum. *Wiley Interdisciplinary Reviews: Water, 6*(5), e1373.
81. National Research Council, Division on Earth, Life Studies, Ocean Studies Board, & Committee on Potential Impacts of Ambient Noise in the Ocean on Marine Mammals. (2003). *Ocean noise and marine mammals*. National Academies Press.
82. Arcangeli, G., Lulli, L. G., Traversini, V., De Sio, S., Cannizzaro, E., Galea, R. P., & Mucci, N. (2022). Neurobehavioral alterations from noise exposure in animals: A systematic review. *International Journal of Environmental Research and Public Health, 20*(1), 591.

83. Guh, Y. J., Tseng, Y. C., & Shao, Y. T. (2021). To cope with a changing aquatic soundscape: Neuroendocrine and antioxidant responses to chronic noise stress in fish. *General and Comparative Endocrinology, 314*, 113918.
84. Duarte, C. M., Chapuis, L., Collin, S. P., Costa, D. P., Devassy, R. P., Eguiluz, V. M., et al. (2021). The soundscape of the Anthropocene ocean. *Science, 371*(6529), eaba4658.
85. Whitehead, A. (2013). Interactions between oil-spill pollutants and natural stressors can compound ecotoxicological effects. *Integrative and Comparative Biology, 53*(4), 635–647.
86. Chang, Y. C., & Saqib, M. (2025). International legal systems in tackling the marine plastic pollution: A critical analysis of UNCLOS and MARPOL. *Water, 17*(10), 1547.
87. Kumar, P., Jose, A. A., & Rajagopal, A. R. (2024). Sustainable use of marine resources using ocean technologies. In *Sustainable development goals* (pp. 249–274). CRC Press.
88. Khan, M. N., & Mohammad, F. (2013). Eutrophication: Challenges and solutions. In *Eutrophication: Causes, consequences and control: Volume 2* (pp. 1–15). Springer Netherlands.
89. Altieri, A. H., & Diaz, R. J. (2019). Dead zones: Oxygen depletion in coastal ecosystems. In *World seas: An environmental evaluation* (pp. 453–473). Academic Press.
90. Ashitha, A., & Rakhimol, K. R. (2021). Fate of the conventional fertilizers in environment. In *Controlled release fertilizers for sustainable agriculture* (pp. 25–39). Academic Press.
91. Wurtsbaugh, W. A., Paerl, H. W., & Dodds, W. K. (2019). Nutrients, eutrophication and harmful algal blooms along the freshwater to marine continuum. *Wiley Interdisciplinary Reviews: Water, 6*(5), e1373. Hasan, B. R., Islam, M. S., Kundu, P., & Mallick, U. K. (2023). Modeling the effects of algal bloom on dissolved oxygen in eutrophic water bodies. *Journal of Mathematics, 2023*(1), 2335570.
92. O'Boyle, S. (2020). Oxygen depletion in coastal waters and the open ocean: Hypoxia and anoxia cases and consequences for biogeochemical cycling and marine life. In *Coastal and deep ocean pollution* (pp. 41–67). CRC Press.
93. Callier, M. D., Byron, C. J., Bengtson, D. A., Cranford, P. J., Cross, S. F., Focken, U., et al. (2018). Attraction and repulsion of mobile wild organisms to finfish and shellfish aquaculture: A review. *Reviews in Aquaculture, 10*(4), 924–949.
94. Aguirre-Velarde, A., Breitburg, D., Cardich, J., Carstensen, J., Dewitte, B., Engel, A., et al. (2021). System controls of coastal and open ocean oxygen depletion. *Progress in Oceanography, 197*, 102613.
95. Dybas, C. L. (2005). Dead zones spreading in world oceans. *BioScience, 55*(7), 552–557.
96. Andreasen, M. H. (2024). *Global change and invasive jellyfish: Hypoxia, fish predation and ecosystem effects*. Technical University of Denmark.
97. Aniebone, V. O., Imo, D. O., & Shonde, O. O. (2024). Nutrients pollution of the coastal marine ecosystems. *International Journal of Innovative Agriculture & Biology Research, 12*(2), 82–95.
98. Vantarakis, A. (2021). Eutrophication and public health. In *Chemical lake restoration: Technologies, innovations and economic perspectives* (pp. 23–47). Springer International Publishing.
99. Anderson, C. R., Sellner, K. G., & Anderson, D. M. (2017). Bloom prevention and control.
100. Hedley, C. (2015). The role of precision agriculture for improved nutrient management on farms. *Journal of the Science of Food and Agriculture, 95*(1), 12–19.
101. Carey, R. O., & Migliaccio, K. W. (2009). Contribution of wastewater treatment plant effluents to nutrient dynamics in aquatic systems: A review. *Environmental Management, 44*(2), 205–217.
102. Jabłońska, E., Winkowska, M., Wiśniewska, M., Geurts, J., Zak, D., & Kotowski, W. (2021). Impact of vegetation harvesting on nutrient removal and plant biomass quality in wetland buffer zones. *Hydrobiologia, 848*(14), 3273–3289.
103. Bohman, B. (2018). Lessons from the regulatory approaches to combat eutrophication in the Baltic Sea region. *Marine Policy, 98*, 227–236.

104. Ye, L., Cai, Q., Zhang, M., & Tan, L. (2014). Real-time observation, early warning and forecasting phytoplankton blooms by integrating in situ automated online sondes and hybrid evolutionary algorithms. *Ecological Informatics, 22*, 44–51.
105. Mashi, S. A., & Shuaibu, H. S. (2018). People and sustainable land management: Assessment of stakeholders knowledge of the nature of landuse/cover change in Abuja, Nigeria. *GeoJournal, 83*(3), 545–562.
106. Altieri, A. H., & Gedan, K. B. (2015). Climate change and dead zones. *Global Change Biology, 21*(4), 1395–1406.
107. Weatherdon, L. V., Magnan, A. K., Rogers, A. D., Sumaila, U. R., & Cheung, W. W. (2016). Observed and projected impacts of climate change on marine fisheries, aquaculture, coastal tourism, and human health: An update. *Frontiers in Marine Science, 3*, 48.
108. Doney, S. C., Busch, D. S., Cooley, S. R., & Kroeker, K. J. (2020). The impacts of ocean acidification on marine ecosystems and reliant human communities. *Annual Review of Environment and Resources, 45*(1), 83–112.
109. Hurd, C. L., Beardall, J., Comeau, S., Cornwall, C. E., Havenhand, J. N., Munday, P. L., et al. (2019). Ocean acidification as a multiple driver: How interactions between changing seawater carbonate parameters affect marine life. *Marine and Freshwater Research, 71*(3), 263–274.
110. Jiang, L. Q., Carter, B. R., Feely, R. A., Lauvset, S. K., & Olsen, A. (2019). Surface ocean pH and buffer capacity: Past, present and future. *Scientific Reports, 9*(1), 18624.
111. Samanth, M. (2024). Ocean acidification and its impact on marine ecosystems. *SSRN 4878195*.
112. Eakin, C. M., Kleypas, J., & Hoegh-Guldberg, O. (2008). Global climate change and coral reefs: Rising temperatures, acidification and the need for resilient reefs. *Status of the Coral Reefs of the World* (pp. 29–34).
113. Martin-Podevin, L. (2023). *The effects of ocean acidification on the olfactory system of fish* (Doctoral dissertation).
114. Mai, R., & Chen, L. (2024). Impacts of ocean acidification on marine ecosystems and mitigation strategies. *International Journal of Marine Science, 14*, 231.
115. Makomere, R. (2024). Responding to ocean acidification by strengthening marine ecosystem health and resilience. In *Research handbook on climate change and biodiversity law* (pp. 162–183). Edward Elgar Publishing.
116. Burke, S. (2024). *Insights into the state of coral reefs and coral health research* (Doctoral dissertation, University of New South Wales (Australia)).
117. Sheppard, C. (2021). *Coral reefs: A very short introduction* (Vol. 391). Oxford University Press.
118. Hughes, T. P., Kerry, J. T., Álvarez-Noriega, M., Álvarez-Romero, J. G., Anderson, K. D., Baird, A. H., et al. (2017). Global warming and recurrent mass bleaching of corals. *Nature, 543*(7645), 373–377.
119. Moriarty, T., Leggat, W., Heron, S. F., Steinberg, R., & Ainsworth, T. D. (2023). Bleaching, mortality and lengthy recovery on the coral reefs of Lord Howe Island. The 2019 marine heatwave suggests an uncertain future for high-latitude ecosystems. *PLoS Climate, 2*(4), e0000080.
120. Roberts, S. (2024). Reef ecosystem conservation for the Hawaiian Islands: The impacts of ocean warming and acidification on benthic reef populations.
121. Good, A. M., & Bahr, K. D. (2021). The coral conservation crisis: Interacting local and global stressors reduce reef resiliency and create challenges for conservation solutions. *SN Applied Sciences, 3*(3), 312.
122. Anthony, K. R. (2016). Coral reefs under climate change and ocean acidification: Challenges and opportunities for management and policy. *Annual Review of Environment and Resources, 41*(1), 59–81.
123. Jha, M. K., & Dev, M. (2024). Impacts of climate change. In *Smart internet of things for environment and healthcare* (pp. 139–159). Springer Nature Switzerland.

124. Church, J. A., Gregory, J. M., Huybrechts, P., Kuhn, M., Lambeck, K., Nhuan, M. T., et al. (2001). Changes in sea level. In J. T. Houghton, Y. Ding, D. J. Griggs, M. Noguer, P. J. Van der Linden, X. Dai, K. Maskell, & C. A. Johnson (Eds.), *Climate change 2001: The scientific basis: Contribution of Working Group I to the third assessment report of the Intergovernmental Panel* (pp. 639–694). Cambridge University Press.
125. Pielke, R., Jr., Burgess, M. G., & Ritchie, J. (2022). Plausible 2005–2050 emissions scenarios project between 2 and 3 degrees C of warming by 2100. *Environmental Research Letters, 17*, 024027.
126. Nichols, C. R., Zinnert, J., & Young, D. R. (2018). Degradation of coastal ecosystems: Causes, impacts and mitigation efforts. In *Tomorrow's coasts: Complex and impermanent* (pp. 119–136). Springer International Publishing.
127. Tully, K., Gedan, K., Epanchin-Niell, R., Strong, A., Bernhardt, E. S., BenDor, T., et al. (2019). The invisible flood: The chemistry, ecology, and social implications of coastal saltwater intrusion. *BioScience, 69*(5), 368–378.
128. Telesh, I., Schubert, H., & Skarlato, S. (2013). Life in the salinity gradient: Discovering mechanisms behind a new biodiversity pattern. *Estuarine, Coastal and Shelf Science, 135*, 317–327.
129. Armstrong, C., & Corbett, J. (2021). Climate change, sea level rise, and maritime baselines: Responding to the plight of low-lying atoll states. *Global Environmental Politics, 21*(1), 89–107.
130. Walsh, K. J., McBride, J. L., Klotzbach, P. J., Balachandran, S., Camargo, S. J., Holland, G., et al. (2016). Tropical cyclones and climate change. *Wiley Interdisciplinary Reviews: Climate Change, 7*(1), 65–89.
131. Islam, F. S. (2025). Assessment of the Global Climatic Impacts due to El Nino and La Nina Events. *Journal of Global Ecology and Environment, 21*(3), 1–26.
132. Santojanni, F. B., Miner, H., Hain, H., & Sutton, G. (2023). The impact of climate change on biodiversity in coastal ecosystems. *Jurnal Ilmu Pendidikan Dan Humaniora, 12*(3), 167–182.
133. Laffoley, D., Baxter, J. M., Day, J. C., Wenzel, L., Bueno, P., & Zischka, K. (2019). Marine protected areas. In *World seas: An environmental evaluation* (pp. 549–569). Academic Press.

Chapter 3
Strategies for Sustainability

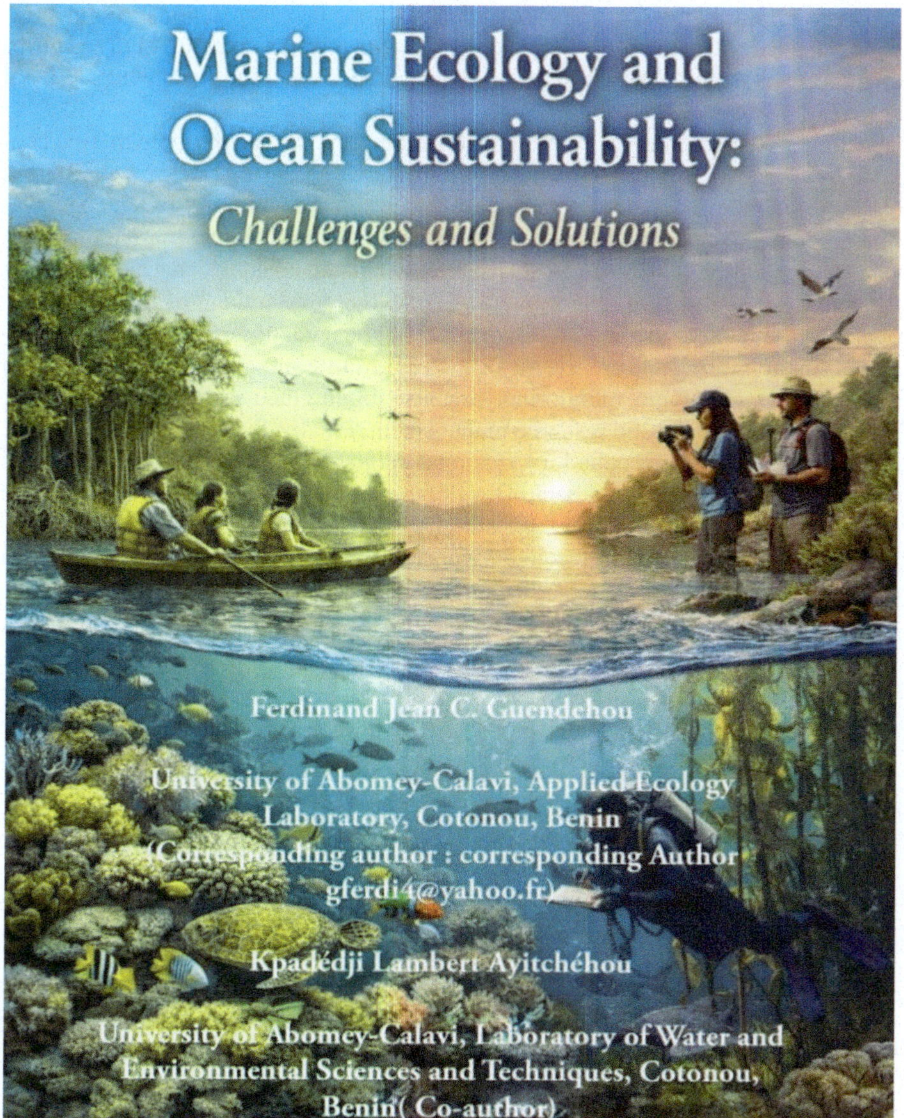

F. J. C. Guendehou, K. L. Ayitchéhou, *Marine Ecology and Ocean Sustainability*, https://doi.org/10.1007/978-3-032-32756-7_3

This research highlights the importance of comprehensive and coordinated strategies to ensure long-term ocean sustainability. It emphasizes the critical role of marine protected areas (MPAs), ecosystem-based management, sustainable fishing practices, and the development of a resilient blue economy. These approaches contribute not only to the conservation but also to climate resilience, food security, and sustainable economic development for coastal communities. Furthermore, effective marine governance relies on strong legal frameworks, international treaties, technological monitoring systems, and the active participation of stakeholders, including local communities, governments, and nongovernmental organizations. By integrating ecological science with policy and governance mechanisms, this study contributes to advancing sustainable ocean management and achieving global environmental goals.

This chapter examines key strategies for promoting sustainability in marine ecosystems. First, it explores the design and effectiveness of marine protected areas (MPAs), including global initiatives, case studies, and community-based conservation models that demonstrate successful approaches to protecting marine biodiversity. This chapter then discusses sustainable management of marine resources through ecosystem-based management (EBM), the development of the blue economy, and the implementation of sustainable fisheries practices. Technological innovations in monitoring and enforcement, such as satellite surveillance, vessel tracking systems, and electronic monitoring, are also highlighted as essential tools for improving compliance and resource management. Finally, the chapter addresses the role of policy, governance, and international cooperation, with particular emphasis on the United Nations Sustainable Development Goal 14 (Life Below Water), the United Nations Convention on the Law of the Sea (UNCLOS), international treaties, and the contributions of nongovernmental organizations in promoting ocean conservation. Together, these strategies illustrate the pathways toward integrated, science-based, and collaborative approaches necessary to ensure the long-term sustainability of the world's oceans.

1 Marine Protected Areas (MPAs) and Conservation

1.1 Design and Effectiveness of Marine Protected Areas (MPAs)

Marine protected areas (MPAs) are designated zones in the ocean aimed at conserving biodiversity, restoring ecosystems, and promoting the use of sustainable resources [1]. Their success depends heavily on thoughtful design, which shapes ecological results and community support [2].

1.1.1 Key Design Principles

1. **Clear Objectives**: MPAs must have specific goals, such as conserving biodiversity, rebuilding fish stocks, or preserving cultural heritage, which guide the zoning, management, and monitoring [3].
2. **Ecological Criteria:** Effective MPAs represent diverse habitats (reefs, mangroves, seagrass, and deep sea), include multiple habitat examples to reduce risk, and support connectivity for species dispersal and migration. Larger and more compact areas tend to perform better, and priority is given to zones with climate resilience [3].
3. **Zoning and Protection Levels**: MPAs often include zones with varying protection:
   (a) *No-take zones* prohibit extraction and provide the greatest ecological benefits.
   (b) *Multiple use zones* allow regulated activities, such as tourism or artisanal fishing.
   (c) *Temporal closures* protect species during breeding or migration periods [4].
4. **Stakeholder Engagement**: Involving fishermen, Indigenous people, scientists, and local communities in the planning process promotes trust, compliance, and incorporates valuable local knowledge [5].
5. **Governance and Enforcement**: Successful MPAs require clear legal frameworks, defined boundaries, adequate funding, trained personnel, and enforcement capacity. Co-management models with local involvement can boost effectiveness [6].
6. **Monitoring and Adaptive Management**: Ongoing monitoring of ecological and socioeconomic indicators allows adaptive responses to changes, such as climate impacts or changes in resource use [7].
7. **Network Design**: MPAs are more effective when they form interconnected networks, which support the resilience of the ecosystem and protect migratory species [8].

Effectiveness Factors MPAs are most effective when they are large, well enforced, long established, and highly or fully protected particularly in no-take zones. Strong community and stakeholder support further enhances their performance [9]. When designed and managed effectively, MPAs improve biodiversity, ecosystem health, and fisheries sustainability through spillover benefits, while boosting resilience against climate change [10].

1.2 Global Initiatives and Case Studies

1.2.1 Global Initiatives on Marine Protected Areas (MPAs)

Marine protected areas (MPAs) are at the forefront of global strategies to protect ocean health, conserve biodiversity, and maintain vital ecosystem services. In response to growing threats, such as overfishing, pollution, habitat loss, and climate change, international efforts have intensified to expand and strengthen MPAs worldwide [11]. One of the most significant frameworks is the Convention on Biological Diversity (CBD), which set the "30 × 30" target through the 2022 Kunming-Montreal Global Biodiversity Framework, committing nations to protect at least 30% of the ocean by 2030 [12]. This goal emphasizes the need for MPAs that are ecologically representative, well connected, and equitably managed. Similarly, the UN Sustainable Development Goal 14 (SDG 14) aims to conserve 10% of coastal and marine areas [13]. While approximately 8–9% of marine areas are currently under protection, there are challenges to ensure enforcement, quality management, and equity [14]. The SDG continues to guide global progress in marine governance. The High Ambition Coalition for Nature and People, comprising more than 100 countries, reinforces the 30 × 30 target by promoting both MPAs and other effective area-based conservation measures (OECMs). This initiative underscores the inclusion of indigenous rights, climate resilience, and biodiversity outcomes [13]. In 2023, the BBNJ Agreement under the United Nations Convention on the Law of the Sea (UNCLOS) was adopted, creating a legal mechanism to establish MPAs in international waters, a critical step for safeguarding the high seas, which span nearly half of the Earth's surface [15]. The Regional Seas Conventions, such as OSPAR, HELCOM, and the Nairobi Convention (under UNEP), support the development of MPA across geopolitical boundaries. These agreements encourage cross-border collaboration, ecosystem-based management, and data sharing [16]. Political alliances such as the Global Ocean Alliance and Blue Leaders advocate for improved ocean protection, improved MPA enforcement, and the advancement of sustainable blue economies [17]. Scientific initiatives, including the identification of Important Marine Mammal Areas (IMMA) and Key Biodiversity Areas (KBA), guide MPA placement by highlighting ecologically significant habitats [18]. To ensure equitable global progress, financial and technical support from international funds and organizations such as the Global Environment Facility (GEF), Blue Nature Alliance, and Pew Bertarelli Ocean Legacy focuses on building capacity and enabling enforcement, especially in developing nations [19]. In conclusion, global initiatives for MPAs are gaining momentum both in scope and ambition. Meeting goals like 30 × 30 requires not only expanding protected areas but also ensuring that they are effectively managed, inclusive of local and indigenous communities, and ecologically resilient. Collectively, these efforts are critical to ensuring a sustainable, thriving ocean for future generations.

1.2.2 Case Studies

Marine protected areas (MPAs) around the world showcase diverse strategies in marine conservation, reflecting varying scales, governance models, and results [20]. The following case studies provide information on successful approaches and ongoing challenges in MPA design and implementation [20].

1. **Great Barrier Reef Marine Park (Australia)**

Spanning approximately 344,000 km^2, this is one of the largest and most iconic MPAs. It uses a multi-use zoning system, with about one-third designated as no-take zones. Its effectiveness is based on strong science-based management, legal frameworks, and stakeholder involvement, supporting biodiversity, fisheries, and tourism. However, climate change, especially coral bleaching, remains a major threat to reef long-term health [21].

2. **Papahānaumokuākea Marine National Monument (USA)**

Covering more than 1.5 million km^2 in the Northwestern Hawaiian Islands, it is among the largest fully protected MPAs worldwide. It combines federal protection with Indigenous Hawaiian knowledge and strict no-take rules. Its remote location, cultural integration, and comprehensive protection contribute to its ecological effectiveness and global significance [22].

3. **Cabo Pulmo National Park (Mexico)**

This small (~71 km^2) MPA in Baja California demonstrates the power of community-led conservation. Once overfished, the reef saw a 400% increase in biomass within a decade after being designated a no-take zone in 1995. Strong local support, scientific monitoring, and effective enforcement turned it into a model for ecological recovery and sustainable tourism [23].

4. **Chagos Marine Protected Area (British Indian Ocean Territory)**

Established in 2010, this ~640,000 km^2 no-take MPA is ecologically important but politically controversial. Its creation excluded the displaced Chagossian people, sparking an international debate over human rights and the colonial legacy. Although it protects biodiversity, its legitimacy is questioned, highlighting the need for inclusive and socially just conservation practices [24].

5. **Locally Managed Marine Areas (LMMAs) in Fiji**

Fiji hosts more than 400 LMMAs, where local communities manage coastal resources using a blend of traditional knowledge and modern conservation methods. These MPAs improve habitat protection, support sustainable fisheries, and empower local governance. Their success is based on strong community leadership, customary laws, and bottom-up management [25]. ***In conclusion***, these case studies reveal that effective MPAs can exist on various scales from massive no-take zones to small community-managed areas. Key success factors include strong legal backing, scientific support, stakeholder and Indigenous participation, and adaptive governance. However, challenges such as climate change, social justice, and enforcement persist. Ultimately, inclusive, context-specific approaches are essential to build resilient and equitable marine protection worldwide.

1.3 Community-Based Conservation in Marine Protected Areas (MPAs)

Community-based conservation in marine protected areas (MPAs) empowers local and indigenous communities to take the lead in planning, managing, and safeguarding marine environments. This approach acknowledges that those who depend most on the ocean for their livelihoods, culture, and identity are often better equipped to protect it effectively [26].

1.3.1 Core Principles

At the heart of community-based MPAs is local leadership, where communities initiate or co-manage areas using traditional knowledge and stewardship practices. This grassroots involvement improves compliance, relevance, and legitimacy [27]. Participatory planning ensures that all voices of fishermen, elders, women, and youth are involved in decision-making, helping align conservation goals with social and cultural needs [28]. Many of these MPAs operate under customary governance systems, which include long-standing conservation practices, such as seasonal closures and taboo zones [29]. Through adaptive management, communities monitor ecological and social outcomes and adjust strategies as conditions change, promoting resilience and sustainability [30].

1.3.2 Benefits: Community-Managed

MPAs have demonstrated significant ecological gains, such as increased fish biomass and coral recovery. They contribute to the security through sustainable fishing and ecotourism, which generate income while preserving resources [31]. These initiatives also support cultural revival, reinforcing traditional practices and connections to the sea. Furthermore, they promote stronger governance, build local institutions, and reduce dependence on external enforcement mechanisms [32].

1.3.3 Examples of Success

In Fiji, more than 400 locally managed marine areas (LMMAs) are based on traditional qoliqoli (fishing) rights, protecting reefs and supporting coastal livelihoods [33]. In Madagascar, the Velondriake LMMA combines no-take zones, regulated octopus fishing, and ecotourism, all led by local fishermen with NGO support [34]. The Apo Island MPA in the Philippines, established in the 1980s, is a global model, where no-take zones have led to increased fish stocks and tourism benefits [35].

1.3.4 Challenges

Despite their success, community-based MPAs face several obstacles. Funding and technical support are often limited, restricting monitoring and enforcement. Many lack legal recognition, making integration into national policies difficult [36]. They are also vulnerable to external pressures, such as climate change, industrial fishing, and coastal development, which can undermine local efforts [37].

In conclusion, community-based MPAs are powerful, inclusive tools for marine conservation. When provided with legal support, scientific guidance, and long-term investment, they offer effective, equitable, and culturally grounded models for ocean protection anchored in local leadership and traditional ecological knowledge.

2 Sustainable Marine Resource Management

2.1 Ecosystem-Based Management (EBM)

Ecosystem-based management (EBM) is a holistic approach to sustainable marine resource management that recognizes the full spectrum of interactions within an ecosystem, including humans, rather than focusing on a single species, sector, or issue [38]. Unlike traditional management that targets individual fisheries or activities in isolation, EBM considers cumulative impacts, ecological relationships, and the need to maintain ecosystem services [39]. The key principles of EBM include the maintenance of ecosystem integrity and resilience, the integration of ecological, social, economic, and institutional perspectives, and the management on spatial and temporal scales [39]. It emphasizes adaptive management informed by scientific data, stakeholder participation, and a precautionary approach in the face of uncertainty. In marine contexts, EBM involves regulating fisheries, tourism, coastal development, and pollution in a coordinated manner [40]. For example, managing fish stocks with EBM includes not only setting catch limits but also preserving habitat, considering predator–prey dynamics, and minimizing bycatch and habitat destruction. It often relies on marine spatial planning (MSP), the use of marine protected areas (MPAs), and ecosystem modeling to guide decisions [41]. The approach also addresses the impacts of climate change, ocean acidification, and biodiversity loss by promoting resilience through habitat protection, ecosystem connectivity, and sustainable livelihoods for coastal communities [42]. Effective EBM requires governance frameworks that are participatory, transparent, and capable of enforcing integrated policies between jurisdictions [39]. Ultimately, EBM seeks to maintain long-term ecological health and the benefits humans derive from the ocean, such as food, climate regulation, cultural value, and recreation, by aligning human use with the natural dynamics and limits of marine ecosystems [40].

2.2 *Blue Economy and Sustainable Fisheries*

2.2.1 Blue Economy

The Blue Economy is a framework for sustainable management of marine resources that promotes economic growth, improved livelihoods, and the health of the ocean ecosystem. It emphasizes harnessing the economic potential of oceans, seas, and coasts while ensuring environmental sustainability and social equity [43]. The core sectors include fisheries, aquaculture, maritime transport, coastal tourism, offshore renewable energy, and marine biotechnology. The Blue Economy seeks to balance economic development with ocean conservation, addressing challenges, such as overfishing, pollution, habitat degradation, and climate change [44]. Key principles include sustainable use of marine resources, ecosystem protection, inclusive growth, and innovation. Fisheries and aquaculture are managed to prevent overexploitation and ensure long-term productivity through science-based quotas, ecosystem-based approaches, and certification schemes. Tourism is developed in ways that protect marine environments and benefit local communities [45]. Renewable energy sources such as offshore wind, wave, and tidal power are expanded to reduce carbon emissions while minimizing ecological impacts [46]. The Blue Economy also prioritizes circular economy practices, such as reducing plastic waste and promoting sustainable shipping. It supports marine spatial planning (MSP) to minimize conflicts among ocean users and protect critical habitats [47]. Investment in ocean science, technology, and data is vital to informed decision-making and adaptive governance. Social inclusion and community engagement are central to ensuring that small-scale fishers, coastal populations, and indigenous communities benefit from and participate in marine development [48]. International cooperation and legal frameworks such as the UN Convention on the Law of the Sea (UNCLOS) underpin the governance of shared marine resources. Ultimately, the blue economy integrates environmental management, economic efficiency, and social equity to create a resilient and sustainable ocean-based economy that benefits current and future generations [49].

2.2.2 Sustainable Fisheries

Sustainable fishing is a core component of marine resource management, aiming to harvest aquatic species at a rate that maintains ecosystem balance, biodiversity, and long-term productivity. This approach ensures that fish populations are healthy, ecosystems are resilient, and fishing communities can sustain their livelihoods over time [50]. The key principles of sustainable fisheries include preventing overfishing, protecting habitats, reducing bycatch, and supporting ecosystem-based management. Science-based quotas and total allowable catches (TACs) are established to align harvest rates with the reproductive capacity of fish stocks [41]. Monitoring, control, and surveillance (MCS) systems, along with traceability, are essential to enforce compliance and combat illegal, unregulated, and unregulated (IUU) fishing

[51]. Selective fishing gear and techniques help minimize bycatch and habitat destruction, while closed seasons, marine protected areas (MPAs), and no-take zones support stock recovery and ecosystem health [52]. Small-scale and artisanal fisheries, which often rely on traditional knowledge and local stewardship, play a vital role and are increasingly recognized in policy frameworks for their potential for sustainability [53]. Sustainable aquaculture complements wild capture by reducing pressure on natural stocks, although it requires responsible management to prevent pollution, disease, and invasive species. Certification schemes (e.g., MSC and ASC) and eco-labels promote market-based incentives for sustainable practices [54]. Effective governance is critical and requires cooperation across national and regional levels, especially for shared and migratory stocks. Institutions such as Regional Fisheries Management Organizations (RFMOs) and frameworks such as the FAO Code of Conduct for Responsible Fisheries guide global efforts [55]. Ultimately, sustainable fishing balances ecological integrity with economic and social needs, supporting food security, livelihoods, and marine biodiversity. They are fundamental to a resilient Blue Economy and the broader goal of sustainable ocean governance [56].

2.3 *Technological Innovations in Monitoring and Enforcement*

2.3.1 Technological Innovations in Monitoring

Technological innovations in monitoring are transforming sustainable marine resource management by improving data collection, compliance, and ecosystem understanding. These tools enable real-time, cost-effective tracking of human activities, marine species, and environmental conditions, supporting informed, adaptive decision-making [57]. Satellite technologies and remote sensing provide near-real-time wide-scale data on ocean temperature, chlorophyll levels, pollution, vessel movements, and illegal fishing [58]. Automatic Identification Systems (AIS) and Vessel Monitoring Systems (VMS) track fishing vessel positions, speeds, and routes, helping enforce regulations and detect unauthorized activities [59]. Electronic monitoring (EM), including cameras and sensors, records fishing operations to verify catches, bycatch, and compliance with quotas and gear restrictions [60]. Combined with electronic logbooks and machine learning, EM improves transparency and data accuracy without constant human observers. Drones and autonomous vehicles (AUVs and ROVs) conduct underwater surveys, map seafloor habitats, and monitor marine life in remote or sensitive areas, reducing cost and risk [57]. Acoustic technologies like sonar and passive acoustic monitoring help assess fish abundance and detect species, such as whales or illegal dynamite fishing. DNA-based tools, including environmental DNA (eDNA), allow noninvasive detection of species presence and biodiversity by analyzing water samples [61]. These are particularly valuable for tracking endangered or elusive species and detecting invasive organisms early. Artificial intelligence (AI) and big data analytics synthesize massive data sets from various sources to model ecosystems, predict fish stock dynamics, and optimize management strategies [62]. Blockchain enhances traceability in seafood supply chains, ensuring sustainable sourcing and reducing fraud. These technologies are increasingly integrated into marine spatial planning (MSP), marine protected area (MPA) enforcement, and ecosystem-based management (EBM) [63]. When combined with community-based monitoring and inclusive governance, they help build trust and equity [64]. Ultimately, technological innovation strengthens the ability to manage marine resources sustainably, making ocean governance more precise, accountable, and responsive to environmental change [65].

2.3.2 Enforcement

Enforcement is a critical pillar of sustainable marine resource management, ensuring that regulations designed to protect marine ecosystems and sustain fish stocks are implemented effectively [66]. Without strong enforcement, overfishing, illegal activities, and habitat degradation can undermine even the best management plans. Enforcement involves monitoring, surveillance, compliance, and legal action to uphold fisheries laws, marine protected areas (MPAs), quotas, gear restrictions, and closed seasons [67]. Key tools include patrol boats, satellite tracking, aerial surveillance, and port inspections. Technologies such as Vessel Monitoring Systems (VMS), Automatic Identification Systems (AIS), and Electronic Monitoring (EM) provide real-time data on vessel activities, helping to detect violations, such as illegal, unreported, and unregulated (IUU) fishing [68]. International cooperation is essential, especially in shared or high-sea fisheries. Regional Fisheries Management Organizations (RFMOs), along with treaties such as the Port State Measures Agreement (PSMA), coordinate transboundary enforcement and information sharing [69]. Coastal states, meanwhile, must build capacity for patrols, data analysis, and legal prosecution. Community-based enforcement complements formal systems by involving local fishermen and indigenous groups in surveillance and reporting. This participatory approach promotes stewardship, builds trust, and improves compliance, particularly in small-scale and remote fisheries [70]. The legal framework must include clear penalties, judicial processes, and political will to prosecute offenders. Transparency and public access to enforcement data strengthen accountability. Technological innovations, such as drones, artificial intelligence, and mobile reporting applications, enhance enforcement in challenging environments with limited resources. Corruption, weak governance, and lack of funding often hinder enforcement efforts [71]. Addressing these requires institutional strengthening, cross-sector collaboration, and sustained investment in capacity building. Effective enforcement is not only punitive, it also supports education, capacity development, and incentives for compliance [72]. When integrated with monitoring, science-based management, and stakeholder engagement, robust enforcement ensures that marine resources are used responsibly, ecosystems remain healthy, and coastal communities thrive [72].

3 Policy, Governance, and International Cooperation

3.1 UN Sustainable Development Goal 14: Life Below Water

UN Sustainable Development Goal 14 (SDG 14): *Life Below Water* is a central pillar for global efforts to protect marine ecosystems, ensure sustainable ocean use, and foster international cooperation. It aims to "conserve and sustainably use the oceans, seas and marine resources for sustainable development," recognizing the vital role in climate regulation, food security, biodiversity, and economic livelihoods [73]. The goal emphasizes integrated ecosystem-based approaches to marine management, aligning closely with marine spatial planning (MSP), sustainable fisheries, marine protected areas (MPAs), and climate resilience strategies [74]. SDG 14 encourages cooperation among nations, especially in managing shared and high seas resources, combating illegal, unreported, and unregulated (IUU) fishing, and supporting the capacity of developing countries through financial and technical aid [75]. Progress in SDG 14 is measured through indicators such as the proportion of sustainable fish stocks, the coverage of key marine biodiversity areas, and the degree of ocean acidification [76]. Despite commitments, implementation remains uneven due to governance gaps, lack of enforcement, and limited funding, particularly in the Global South. Achieving SDG 14 requires strong global governance, regional collaboration, and inclusive participation of governments, industries, scientists, and communities [77]. It calls for cross-sector alignment linking marine sustainability to other SDGs, such as poverty reduction, climate action, and responsible consumption [71]. Ultimately, SDG 14 is a blueprint for balancing ocean conservation with human development, advancing ocean stewardship through policy coherence, international cooperation, and science-informed governance to ensure the health and productivity of marine ecosystems for generations to come [78].

3.2 Law of the Sea and International Treaties

3.2.1 The United Nations Convention on the Law of the Sea (UNCLOS)

The United Nations Convention on the Law of the Sea (UNCLOS), often referred to as the "Constitution of the Oceans," is the foundational legal framework that governs all aspects of ocean space, marine resource use, and maritime conduct. Adopted in 1982 and in force since 1994, UNCLOS establishes a comprehensive legal order for the seas and is essential for sustainable marine governance and international cooperation [79]. UNCLOS defines maritime zones that include territorial seas (up to 12 nautical miles), exclusive economic zones (EEZs, up to 200 nautical miles), and the high seas. Within their EEZs, coastal states have sovereign rights to exploit, conserve, and manage marine resources, but must do so responsibly and according to international law [80]. Beyond national jurisdictions, the ocean floor and its resources designated as "the common heritage of mankind" are managed through the International Seabed Authority (ISA). A key element of UNCLOS is the duty to protect and preserve the marine environment [81]. States are required to prevent, reduce, and control pollution; protect marine biodiversity; and conduct environmental impact assessments (EIA) for activities that may cause significant harm [82]. It also promotes marine scientific research, technology transfer, and the development of capacity in developing countries [83]. UNCLOS establishes mechanisms for peaceful resolution of disputes through institutions, such as the International Tribunal for the Law of the Sea (ITLOS), the International Court of Justice (ICJ), and arbitration processes [84]. It underpins cooperation on issues such as fisheries management, marine pollution control, and freedom of navigation, which are essential for global ocean sustainability [85]. Although UNCLOS provides a robust legal foundation, evolving challenges such as climate change, deep-sea mining, marine genetic resources, and biodiversity beyond national jurisdiction (BBNJ) have exposed governance gaps [86]. These are being addressed through new agreements, such as the 2023 BBNJ Treaty under UNCLOS. In essence, UNCLOS is the legal backbone of marine policy and governance. Its principles shape national laws, regional agreements, and international efforts to achieve ocean sustainability, balancing state sovereignty with collective responsibility to protect life below water [87].

3.2.2 International Treaties

International treaties are essential tools in global marine governance, providing legally binding frameworks for cooperation, conservation, and sustainable use of ocean resources. They help to harmonize national actions, resolve transboundary issues, and establish standards for responsible management of the shared marine environment [88]. At the heart of this is the United Nations Convention on the Law of the Sea (UNCLOS), which defines maritime zones, rights, and responsibilities of states and sets obligations for marine environmental protection, resource management, and dispute resolution. UNCLOS is complemented by sector-specific and regional treaties that address emerging and complex ocean challenges [89]. The Convention on Biological Diversity (CBD) supports the conservation of marine biodiversity through initiatives like the Aichi Biodiversity Targets and the post 2020 Global Biodiversity Framework. Its protocols and working groups promote the sustainable use of marine genetic resources and ecosystem-based approaches to marine protection [90]. The FAO Agreement on Port State Measures (PSMA) targets illegal, unreported, and unregulated (IUU) fishing by improving inspection and compliance protocols at ports. It is reinforced by the FAO Code of Conduct for Responsible Fisheries and agreements with Regional Fisheries Management Organizations (RFMOs), which govern the sustainable use of migratory and shared fish stocks [91]. The Convention on the Prevention of Marine Pollution (MARPOL) addresses pollution from ships, including oil, garbage, and harmful substances. Similarly, the London Convention and Protocol regulate ocean dumping to protect marine ecosystems [92]. The newly adopted High Seas Treaty (BBNJ—Biodiversity Beyond National Jurisdiction) under UNCLOS aims to fill governance gaps in international waters by enabling the creation of marine protected areas (MPAs), regulating marine genetic resources, and strengthening environmental impact assessments and benefit sharing mechanisms [93].

Other significant treaties include the RAMSAR Convention (wetlands protection), CITES (trade in endangered species), and regional sea agreements under UNEP, which facilitate cooperation among coastal states in the management of specific marine regions [94]. Together, these treaties form a complex yet coordinated legal architecture that enables multilateral action for ocean sustainability. Their success depends on national implementation, enforcement, adequate financing, technology transfer, and inclusive participation, especially of developing states and local communities [95]. Strengthened treaty synergy is key to addressing climate change, loss of biodiversity, and depletion in the marine realm [96].

3.3 *Role of NGOs and Regional Agreements*

3.3.1 Role of NGOs

Nongovernmental organizations (NGOs) play a pivotal role in advancing marine ecology and ocean sustainability through advocacy, research, education, policy influence, and community engagement. Operating at the local, national, and international levels, NGOs act as vital intermediaries between governments, scientific institutions, industry, and the public, often filling gaps in governance, capacity, and enforcement [97]. NGOs contribute to the development and advocacy by shaping legislation, promoting environmental standards, and participating in international negotiations such as UN climate talks, Convention on Biological Diversity (CBD) meetings, and the development of the High Seas Treaty. Their presence ensures transparency, accountability, and the inclusion of environmental and social priorities in marine governance [98]. In scientific research and monitoring, NGOs conduct field studies, biodiversity assessments, and ecosystem health assessments, often in collaboration with academic institutions. They generate critical data for marine protected area (MPA) designation, sustainable fisheries management, and climate resilience planning, especially in regions with limited governmental capacity [99]. NGOs also promote public awareness and education, using campaigns, media, and community outreach to inform and engage citizens on marine issues, such as overfishing, plastic pollution, coral bleaching, and habitat loss. Through storytelling and citizen science programs, they foster ocean stewardship and behavior change [100]. At the grassroots level, many NGOs support community-based conservation, empowering coastal populations and indigenous communities to manage local marine resources through co-management models, traditional ecological knowledge, and livelihood alternatives [101]. These efforts improve equity, resilience, and sustainability. NGOs are often at the forefront of legal action and enforcement support, working to expose environmental crimes, challenge harmful industrial practices, and push for stronger compliance with international and national laws. They also provide technical and financial support to conservation initiatives, particularly in developing countries [102]. NGOs such as WWF, Oceana, Greenpeace, Conservation International, and local marine-focused groups play diverse roles across sectors and scales. Despite their influence, there remain challenges, including funding restrictions, political resistance, and the need for deeper collaboration with governments and private stakeholders [103]. In summary, NGOs are essential actors in marine sustainability, driving innovation, accountability, and inclusive governance. Their diverse roles strengthen efforts to protect marine ecosystems, promote equitable use of resources, and achieve long-term ocean health [97] (Figs. 3.1, 3.2 and 3.3).

Fig. 3.1 Mangrove forests as blue carbon ecosystems. Mangroves store significant amounts of carbon and protect coastal environments. **Source image**: NASA Earth Observatory. **Reference**: Duarte et al. (2013). The role of coastal plant communities in the mitigation of climate change. Nature Climate Change

Fig. 3.2 Seagrass meadows as critical marine carbon sinks. Seagrasses contribute significantly to carbon sequestration and marine biodiversity. **Source image**: Blue-Carbon Initiative. **Reference**: McLeod, E. et al. (2011). A blueprint for blue carbon. Frontiers in Ecology and the Environment

Fig. 3.3 Mangrove root systems that support marine biodiversity. Complex root structures provide nursery habitats for fish and crustaceans. **Source image**: Conservation International Image Library. **Reference**: Alongi, D.M. (2014). Carbon cycling and storage in mangrove forests. Annual Review of Marine Science

References

1. Van der Ryn, S., & Cowan, S. (2013). *Ecological design*. Island Press.
2. Grorud-Colvert, K., Sullivan-Stack, J., Roberts, C., Constant, V., Horta e Costa, B., Pike, E. P., et al. (2021). The MPA Guide: A framework to achieve global goals for the ocean. *Science, 373*(6560), eabf0861.
3. Laffoley, D., Baxter, J. M., Day, J. C., Wenzel, L., Bueno, P., & Zischka, K. (2019). Marine protected areas. In *World seas: An environmental evaluation* (pp. 549–569). Academic Press.
4. Kirkman, H. (2013). Choosing boundaries to marine protected areas and zoning the MPAs for restricted use and management. *Ocean and Coastal Management, 81*, 38–48.
5. Thornton, T. F., & Scheer, A. M. (2012). Collaborative engagement of local and traditional knowledge and science in marine environments: A review. *Ecology and Society, 17*(3).
6. Siddique, M. (2014). *Co-learning in Marine Protected Area Management* (Doctoral dissertation, University of the Sunshine Coast).
7. O'Connell, E. (2017). Towards adaptation of water resource systems to climatic and socio-economic change. *Water Resources Management, 31*(10), 2965–2984.
8. Carr, M. H., Robinson, S. P., Wahle, C., Davis, G., Kroll, S., Murray, S., et al. (2017). The central importance of ecological spatial connectivity to effective coastal marine protected areas and to meeting the challenges of climate change in the marine environment. *Aquatic Conservation: Marine and Freshwater Ecosystems, 27*, 6–29.
9. Wells, S., Addison, P. F., Bueno, P. A., Costantini, M., Fontaine, A., Germain, L., et al. (2016). Using the IUCN Green List of Protected and Conserved Areas to promote conservation impact through marine protected areas. *Aquatic Conservation: Marine and Freshwater Ecosystems, 26*, 24–44.
10. McLeod, E., Salm, R., Green, A., & Almany, J. (2009). Designing marine protected area networks to address the impacts of climate change. *Frontiers in Ecology and the Environment, 7*(7), 362–370.
11. Villasante, S., Ainsworth, G. B., Pita, P., Belgrano, A., Bennett, N., & Sumaila, U. R. (2023). The role of marine protected areas (MPAs) in providing ecosystem services to improve ocean and human health. In *Oceans and human health* (pp. 23–37). Academic Press.

12. Aubert, G., & Dudley, N. (2024). Progress on implementing the Kunming-Montreal global biodiversity framework. *European Parliament, Directorate-General for Internal Policies of the Union, 10*, 764564.
13. Diz, D., Johnson, D., Riddell, M., Rees, S., Battle, J., Gjerde, K., et al. (2018). Mainstreaming marine biodiversity into the SDGs: The role of other effective area-based conservation measures (SDG 14.5). *Marine Policy, 93*, 251–261.
14. De Santo, E. M. (2013). Missing marine protected area (MPA) targets: How the push for quantity over quality undermines sustainability and social justice. *Journal of Environmental Management, 124*, 137–146.
15. Assembly, U. G. (2024). Agreement under the United Nations Convention on the Law of the Sea on the Conservation and Sustainable Use of Marine Biological Diversity of Areas beyond National Jurisdiction: Resolution/adopted by the General Assembly.
16. Hamukuaya, H. (2024). Assessment of transboundary environmental issues affecting biodiversity in shared marine ecosystems: Towards formulating harmonised regional framework for conservation of aquatic biodiversity and joint action plan.
17. Hammond, A., & Jones, P. J. (2021). Protecting the 'blue heart of the planet': Strengthening the governance framework for marine protected areas beyond national jurisdiction. *Marine Policy, 127*, 104260.
18. Corrigan, C. M., Ardron, J. A., Comeros-Raynal, M. T., Hoyt, E., Notarbartolo Di Sciara, G., & Carpenter, K. E. (2014). Developing important marine mammal area criteria: Learning from ecologically or biologically significant areas and key biodiversity areas. *Aquatic Conservation: Marine and Freshwater Ecosystems, 24*(S2), 166–183.
19. Kennedy, T. J. (2024). Advancing ocean sustainability for climate-resilient change. In *Globalization and sustainability—Ecological, social and cultural perspectives* (p. 7). IntechOpen.
20. Fulton, E. A., Bax, N. J., Bustamante, R. H., Dambacher, J. M., Dichmont, C., Dunstan, P. K., et al. (2015). Modelling marine protected areas: Insights and hurdles. *Philosophical Transactions of the Royal Society, B: Biological Sciences, 370*(1681), 20140278.
21. Day, J. (2016). The great barrier reef Marine Park—The grandfather of modern MPAs. In *Big, Bold and Blue: Lessons from Australia's marine protected areas* (pp. 65–97). CSIRO Publishing.
22. Kikiloi, K., Friedlander, A. M., Wilhelm, A., Lewis, N. A., Quiocho, K., 'Āila W., Jr., & Kaho'ohalahala, S. (2017). Papahānaumokuākea: Integrating culture in the design and management of one of the world's largest marine protected areas. *Coastal Management, 45*(6), 436–451.
23. Knowlton, N. (2021). Ocean optimism: Moving beyond the obituaries in marine conservation. *Annual Review of Marine Science, 13*(1), 479–499.
24. Beeler, C. S. (2025). Managing the commons: How local participation in decision-making impacts the success of large-scale marine protected areas.
25. Govan, H., Tawake, A., Tabunakawai, K., Jenkins, A., Lasgorceix, A., Schwarz, A. M., et al. (2009). Status and potential of locally-managed marine areas in the South Pacific: Meeting nature conservation and sustainable livelihood targets through wide-spread implementation of LMMAs: Study report.
26. Collier, C. E. (2020). Enabling conditions for community-based comanagement of marine protected areas in the United States. *Marine Policy, 122*, 104244.
27. Chukwuka, A. V., Adegboyegun, A. D., & Adeogun, A. O. (2025). Marine protected areas of Africa: Current challenges and future opportunities in conservation. *Aquatic Sciences, 87*(3), 61.
28. Sithole, S. S., Walters, G. M., Mbatha, P., & Matose, F. (2024). Youth engagement in global conservation governance. *Conservation Biology, 38*(6), e14387.
29. Love, M. W. (2021). Tabu, MPA and community: Nomenclature and the political ecology of marine conservation in Vanuatu. *Journal of Political Ecology, 28*(1), 91–110.

30. Tompkins, E. L., & Adger, W. N. (2004). Does adaptive management of natural resources enhance resilience to climate change? *Ecology and Society, 9*(2).
31. Hargreaves-Allen, V. (2010). *Economic values, distributional impacts and conservation outcomes for coral reef marine protected areas* (Doctoral dissertation, Division of Biology, Imperial College London).
32. Carrasco-Bahamonde, D., Casellas, A., & Araos, F. (2025). Getting our sea back: Indigenous governance and biocultural conservation of coastal and marine commons. *Marine Policy, 178*, 106705.
33. Fache, E., & Breckwoldt, A. (2018). Small-scale managed marine areas over time: Developments and challenges in a local Fijian reef fishery. *Journal of Environmental Management, 220*, 253–265.
34. Price, P., & Boulle, T. (2023). *Livelihood opportunities through scaling marine and coastal EbA*. South African Institute of International Affairs.
35. Alcala, A. C., & Russ, G. R. (2006). No-take marine reserves and reef fisheries management in the Philippines: A new people power revolution. *AMBIO: A Journal of the Human Environment, 35*(5), 245–254.
36. Crawford, B., Kasmidi, M., Korompis, F., & Pollnac, R. B. (2006). Factors influencing progress in establishing community-based marine protected areas in Indonesia. *Coastal Management, 34*(1), 39–64.
37. Fischer, A. P. (2018). Pathways of adaptation to external stressors in coastal natural-resource-dependent communities: Implications for climate change. *World Development, 108*, 235–248.
38. Alexander, K. A., & Haward, M. (2019). The human side of marine ecosystem-based management (EBM): 'Sectoral interplay' as a challenge to implementing EBM. *Marine Policy, 101*, 33–38.
39. O'Hagan, A. M. (2020). Ecosystem-based management (EBM) and ecosystem services in EU law, policy and governance. In *Ecosystem-based management, ecosystem services and aquatic biodiversity: Theory, tools and applications* (pp. 353–372). Springer International Publishing.
40. Anagnostou, C., Katsanevakis, S., Kastanidi, E., Streftaris, N., Pagou, K., & Papathanassiou, E. (2023). Integrated planning for the adaptive management of human activities and supporting marine conservation in the Aegean Sea.
41. Kenny, A. J., Campbell, N., Koen-Alonso, M., Pepin, P., & Diz, D. (2018). Delivering sustainable fisheries through adoption of a risk-based framework as part of an ecosystem approach to fisheries management. *Marine Policy, 93*, 232–240.
42. Santojanni, F. B., Miner, H., Hain, H., & Sutton, G. (2023). The impact of climate change on biodiversity in coastal ecosystems. *Jurnal Ilmu Pendidikan Dan Humaniora, 12*(3), 167–182.
43. Bax, N., Novaglio, C., Maxwell, K. H., Meyers, K., McCann, J., Jennings, S., et al. (2022). Ocean resource use: Building the coastal blue economy. *Reviews in Fish Biology and Fisheries, 32*(1), 189–207.
44. Johnson, K., Masters, I., & Dalton, G. (2018). *Building industries at sea-'blue growth' and the new maritime economy* (p. 514). Taylor & Francis.
45. Hambrey, J. (2017). The 2030 agenda and the sustainable development goals: The challenge for aquaculture development and management. *FAO fisheries and aquaculture circular*, (C1141).
46. Appiott, J., Dhanju, A., & Cicin-Sain, B. (2014). Encouraging renewable energy in the offshore environment. *Ocean and Coastal Management, 90*, 58–64.
47. Elston, J., Pinto, H., & Nogueira, C. (2024). Tides of change for a sustainable blue economy: A systematic literature review of innovation in maritime activities. *Sustainability, 16*(24), 11141.
48. Cvitanovic, C., Hobday, A. J., van Kerkhoff, L., Wilson, S. K., Dobbs, K., & Marshall, N. A. (2015). Improving knowledge exchange among scientists and decision-makers to facilitate the adaptive governance of marine resources: A review of knowledge and research needs. *Ocean and Coastal Management, 112*, 25–35.

49. Simi, K. K. (2023). Marine biodiversity management under the UNCLOS: Scope and challenges in the era of blue economy policy. *GLS Law Journal, 5*(2), 102–112.
50. Roberts, S. J., & Brink, K. (2010). Managing marine resources sustainably. *Environment, 52*(4), 44–52.
51. Yleaña, J. S., & Velasco, P. E. L. (2012). Monitoring, control and surveillance (MCS) in Southeast Asia: Review of the establishment of regional MCS network. *Fish for the People, 10*(1), 7–12.
52. Mehta, H. (2025). *Effective fisheries management through aquatic protected areas*. Educohack Press.
53. McConney, P., Pereira Medeiros, R., Pascual-Fernández, J. J., & Pena, M. (2018). Stewardship and sustainable practices in small-scale fisheries. In *Transdisciplinarity for small-scale fisheries governance: Analysis and practice* (pp. 181–201). Springer International Publishing.
54. Ramachandran, A. (2010). Ecolabeling and green certification for effective fisheries management—An analysis.
55. Ridgeway, L., & Rice, J. C. (2010). International organizations and fisheries governance. In *Handbook of marine fisheries conservation and management* (pp. 485–504). Oxford University Press.
56. Elegbede, I. O., Fakoya, K. A., Adewolu, M. A., Jolaosho, T. L., Adebayo, J. A., Oshodi, E., et al. (2025). Understanding the social–ecological systems of non-state seafood sustainability scheme in the blue economy. *Environment, Development and Sustainability, 27*(2), 2721–2752.
57. Glaviano, F., Esposito, R., Cosmo, A. D., Esposito, F., Gerevini, L., Ria, A., et al. (2022). Management and sustainable exploitation of marine environments through smart monitoring and automation. *Journal of Marine Science and Engineering, 10*(2), 297.
58. Landrigan, P. J., Stegeman, J. J., Fleming, L. E., Allemand, D., Anderson, D. M., Backer, L. C., et al. (2020). Human health and ocean pollution. *Annals of Global Health, 86*(1), 151.
59. Seppä, E. (2024). Evolving Automatic Identification System (AIS) in maritime.
60. van Helmond, A. T., Mortensen, L. O., Plet-Hansen, K. S., Ulrich, C., Needle, C. L., Oesterwind, D., et al. (2020). Electronic monitoring in fisheries: Lessons from global experiences and future opportunities. *Fish and Fisheries, 21*(1), 162–189.
61. Mirone, E. (2024). Non-invasive monitoring of freshwater species in Central and Southern Italian basins. From standard survey to environmental DNA.
62. Ullah, F., Saqib, S., & Xiong, Y. C. (2025). Integrating artificial intelligence in biodiversity conservation: Bridging classical and modern approaches. *Biodiversity and Conservation, 34*(1), 45–65.
63. Patro, P. K., Jayaraman, R., Salah, K., & Yaqoob, I. (2022). Blockchain-based traceability for the fishery supply chain. *IEEE Access, 10*, 81134–81154.
64. Matlala, L. S. (2025). Harnessing citizen-based monitoring for sustainable governance in South Africa: A framework for achieving SDGs through enhanced governance practice. *OIDA International Journal of Sustainable Development, 18*(8), 109–128.
65. Shih, Y. C. (2024). Ocean governance in practice: A study of the application of marine science and technology research techniques to maritime law enforcement in Taiwan. *Marine Policy, 163*, 106081.
66. Pomeroy, R., Parks, J., Reaugh-Flower, K., Guidote, M., Govan, H., & Atkinson, S. (2015). Status and priority capacity needs for local compliance and community-supported enforcement of marine resource rules and regulations in the Coral Triangle Region. *Coastal Management, 43*(3), 301–328.
67. Aboul-Dahab, K. (2022). The role of vessel monitoring systems (VMS) in mitigating illegal, unreported and unregulated (IUU) fishing. *Unreported and Unregulated (IUU) Fishing (April 12, 2022)*.
68. Molenaar, E. J. (2022). International cooperation through regional fisheries management organizations. In *Marine resources, climate change and international management regimes* (p. 23). Bloomsbury Academic.

69. Dao, Y., Yusnaldi, Y., & Kusuma, K. (2024). Mitigating destructive fishing through the optimization of community-based coastal surveillance as an effort to safeguard maritime security. *Formosa Journal of Applied Sciences, 3*(7), 2783–2796.
70. Oluwaferanmi, A. (2025). Bridging innovation and regulation: Legal and technological responses to cryptocurrency-driven financial crime.
71. Tiruwa, A., & Dikshit, S. (2025). Building sustainable futures: Cross-sector partnerships for capacity building under SDG 17. *SSRN 5317795*.
72. Bermas-Atrigenio, N. A., & Chua, T. E. (2013). Utilizing science in advancing marine ecosystem-based management. *Ocean and Coastal Management, 81*, 1–6.
73. Koutouki, K., & Phillips, F. K. (2016). SDG 14 on ensuring conservation and sustainable use of oceans and marine resources: Contributions of International Law, Policy and Governance. *Issue Brief*.
74. Ansong, J., Gissi, E., & Calado, H. (2017). An approach to ecosystem-based management in maritime spatial planning process. *Ocean and Coastal Management, 141*, 65–81.
75. Okafor-Yarwood, I. (2019). Illegal, unreported and unregulated fishing, and the complexities of the sustainable development goals (SDGs) for countries in the Gulf of Guinea. *Marine Policy, 99*, 414–422.
76. Virto, L. R. (2018). A preliminary assessment of the indicators for Sustainable Development Goal (SDG) 14 "Conserve and sustainably use the oceans, seas and marine resources for sustainable development". *Marine Policy, 98*, 47–57.
77. Florini, A., & Pauli, M. (2018). Collaborative governance for the sustainable development goals. *Asia & the Pacific Policy Studies, 5*(3), 583–598.
78. Boteler, B., & Gianni, M. (2023). Atlantic Ocean governance frameworks affecting Atlantic marine ecosystems under conditions of change.
79. Haque, A. E. (2016). *United Nations convention on the law of the sea (UNCLOS) and delimitation of maritime boundaries: A Bangladesh perspective* (Doctoral dissertation, Western Sydney University (Australia)).
80. Treves, T. (2015). Coastal states' rights in the maritime areas under UNCLOS. *Brazilian Journal of International Law, 12*, 40.
81. Franckx, E. (2010). The international seabed authority and the common heritage of mankind: The need for states to establish the outer limits of their continental shelf. *The International Journal of Marine and Coastal Law, 25*(4), 543–567.
82. Elias, O., & Wong, M. (2021). Environmental impact assessment. In *Research handbook on international environmental law* (pp. 188–208). Edward Elgar Publishing.
83. Vierros, M. K., & Harden-Davies, H. (2020). Capacity building and technology transfer for improving governance of marine areas both beyond and within national jurisdiction. *Marine Policy, 122*, 104158.
84. Chowdhury, M. R. K., Hamid, S. A., & Mohd Salleh, N. H. (2024). Maritime dispute resolution: Understanding adequacy of UNCLOS: A systematic review. *Australian Journal of Maritime & Ocean Affairs, 18*, 1–22.
85. Spalding, A. K., & De Ycaza, R. (2020). Navigating shifting regimes of ocean governance: From UNCLOS to sustainable development goal 14. *Environment and Society, 11*(1), 5–26.
86. Humphries, F., Gottlieb, H. M., Laird, S., Wynberg, R., Lawson, C., Rourke, M., et al. (2020). A tiered approach to the marine genetic resource governance framework under the proposed UNCLOS agreement for biodiversity beyond national jurisdiction (BBNJ). *Marine Policy, 122*, 103910.
87. Hutse, R. (2024). *The ship has reached the shore: The role of the UN BBNJ agreement regarding marine protected areas* (Doctoral dissertation, Ghent University).
88. Kimball, L. A. (2003). *International ocean governance: Using international law and organizations to manage marine resources sustainably*. IUCN.
89. Jensen, Ø., & Bankes, N. (2017). Compulsory and binding dispute resolution under the United Nations convention on the law of the sea: Introduction. *Ocean Development & International Law, 48*(3–4), 209–215.

90. Carr, H., Abas, M., Boutahar, L., Caretti, O. N., Chan, W. Y., Chapman, A. S., et al. (2020). The Aichi Biodiversity Targets: Achievements for marine conservation and priorities beyond 2020. *PeerJ, 8*, e9743.
91. Gouache, C. (2021). Proposed guidelines on pre-arrival risk assessments of foreign vessels: Using lessons learned to strengthen implementation of the UN FAO Agreement on Port State Measures.
92. Fitzmaurice, M. (2023). The international convention for the prevention of pollution from ships (MARPOL). In *Research handbook on ocean governance law* (pp. 91–108). Edward Elgar Publishing.
93. Krypotos, P. (2025). *Beyond borders: Assessing the high seas treaty's impact on global marine biodiversity* (Master's thesis, Πανεπιστήμιο Πειραιώς).
94. Dragomir, A. N., & Florescu, I. (2024). Challenges and strategies for implementing the Ramsar Convention. Balancing economic development and wetland conservation. *Studia Ecologiae et Bioethicae, 22*(4), 57–71.
95. Warner, R., & Rayfuse, R. (2008). Securing a sustainable future for the oceans beyond national jurisdiction: The legal basis for an integrated cross-sectoral regime for high seas governance for the 21st century. *The International Journal of Marine and Coastal Law, 23*(3), 399–421.
96. Zreik, M. (2024). Synergizing trade and sustainability: Advancing SDG 14 through international trade dynamics. *Marine Development, 2*(1), 13.
97. Crosman, K. (2013). *The role of non-governmental organizations in marine conservation* (Doctoral dissertation).
98. Kenney, N. (2024). International human rights law in the context of the convention on biological diversity's financial mechanism: The responsibility of the global environment facility's implementing agencies to respect, protect and fulfil human rights in biodiversity-related financial assistance.
99. Fox, H. E., Barnes, M. D., Ahmadia, G. N., Kao, G., Glew, L., Haisfield, K., et al. (2017). Generating actionable data for evidence-based conservation: The global center of marine biodiversity as a case study. *Biological Conservation, 210*, 299–309.
100. Perez, L. (2017). *Marine litter: The plastic pollution in the ocean crisis and how to raise public awareness to impact legislation* (Master's thesis, University of Southern California).
101. Caldeira, M., Sekinairai, A. T., & Vierros, M. (2025). Weaving science and traditional knowledge: Toward sustainable solutions for ocean management. *Marine Policy, 174*, 106591.
102. White, R. (2012). NGO engagement in environmental law enforcement: Critical reflections. *Australasian Policing, 4*(1), 4–7.
103. Weatherley, A. (2024). An interdisciplinary exploration of deep-sea mining in Papua New Guinea.

Chapter 4
Future Directions and Solutions

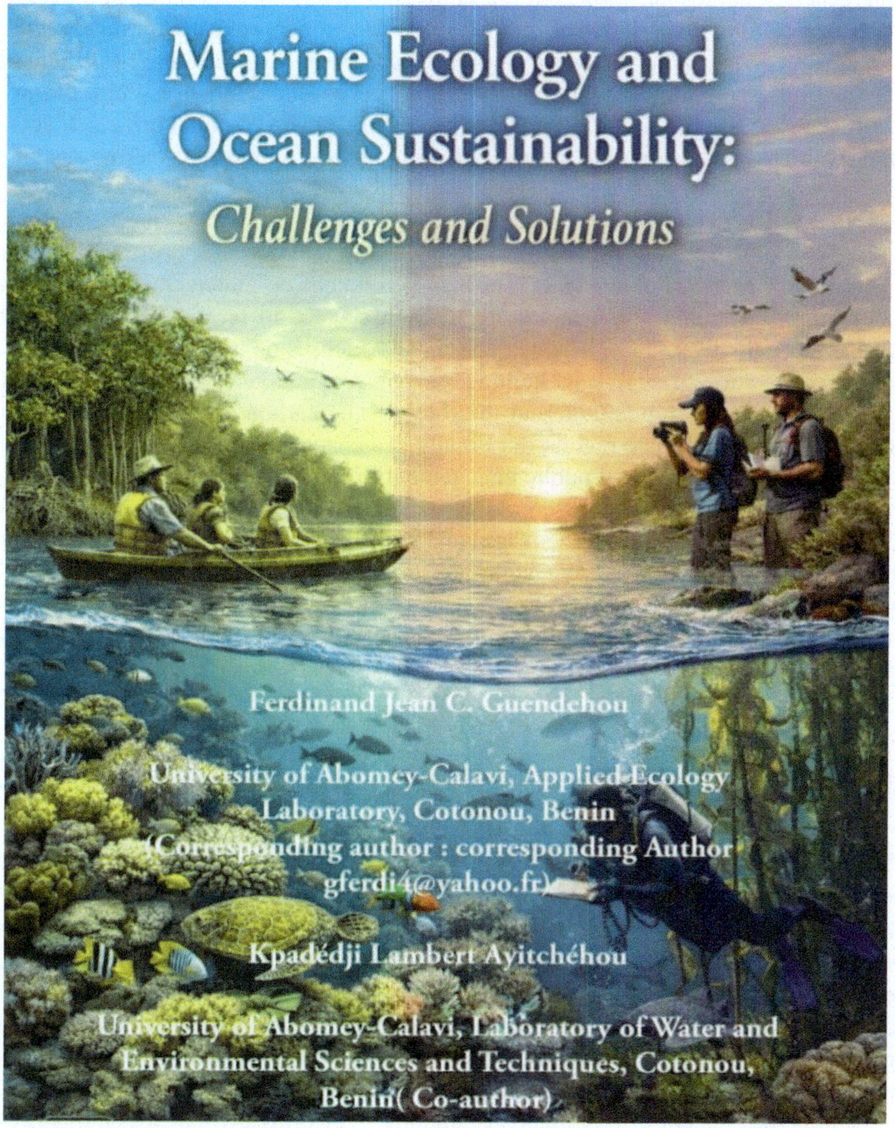

F. J. C. Guendehou, K. L. Ayitchéhou, *Marine Ecology and Ocean Sustainability*, https://doi.org/10.1007/978-3-032-32756-7_4

This research highlights the importance of integrating traditional ecological knowledge, scientific innovation, and community participation in shaping future marine conservation strategies. Indigenous marine stewardship and co-management frameworks demonstrate that sustainable ocean governance can benefit from culturally grounded knowledge systems that emphasize long-term ecological balance and intergenerational responsibility. At the same time, emerging restoration approaches such as coral reef rehabilitation, kelp forest recovery, mangrove replanting, and the protection of blue carbon ecosystems offer nature-based solutions that contribute to climate mitigation, biodiversity conservation, and coastal resilience. Furthermore, education, media participation, and youth participation are essential drivers of societal awareness and behavioral change, allowing broader support for marine sustainability policies and practices.

This section explores future directions and solutions to advance marine sustainability through integrated ecological, social, and governance approaches. It first examines the role of indigenous marine stewardship and co-management systems in combining traditional knowledge with modern science to promote inclusive and adaptive ocean governance. The discussion then focuses on marine restoration and climate adaptation strategies, including coral reef restoration techniques, kelp forest recovery, mangrove replanting, and the role of blue carbon ecosystems in carbon sequestration and climate regulation. Finally, the chapter highlights the importance of education, advocacy, and public engagement in fostering ocean literacy and sustainable behavior. Particular attention is paid to awareness campaigns, behavioral change strategies, the role of the media and citizen science, and youth engagement initiatives that empower future generations to participate actively in marine conservation. Together, these approaches illustrate the pathways toward resilient, inclusive, and science-based solutions capable of ensuring the long-term sustainability of ocean ecosystems and the communities that depend on them.

1 Integrating Traditional Knowledge and Science

1.1 Indigenous Marine Stewardship

Indigenous marine stewardship represents a vital and evolving approach to ocean sustainability, rooted in traditional knowledge systems that have managed marine environments for millennia [1]. Indigenous communities around the world have deep ecological understanding, spiritual relationships with the sea, and place-based management practices that align closely with modern principles of conservation, resilience, and ecosystem-based management [2]. At the heart of indigenous stewardship is the concept of relational sustainability, a worldview that sees humans as part of, rather than separate from, marine ecosystems [3]. Practices such as seasonal harvesting, sacred no-take zones, taboos, and transmission of knowledge through oral histories have long maintained ecological balance and protected biodiversity in coastal and marine areas [4]. Indigenous governance systems often emphasize intergenerational responsibility, holistic observation, and adaptive responses to environmental change [5]. These systems have proven especially resilient in the face of climate impacts and resource pressures, making them essential for future marine conservation and sustainability planning [6]. Today, Indigenous marine stewardship is gaining global recognition through co-management frameworks, legal rights to marine territories, and policy integration. Examples include First Nations Guardian Programs in Canada, the Rhui system in Polynesia, and the Sea Country plans of Aboriginal groups [7]. These initiatives combine traditional knowledge with scientific research, satellite monitoring, and ecological surveys to manage fisheries, restore habitats, and protect sacred marine spaces [8]. Recognition of Indigenous and local knowledge (ILK) in international frameworks, such as the Convention on Biological Diversity (CBD) and the Intergovernmental Science-Policy Platform on Biodiversity and Ecosystem Services (IPBES), is driving efforts to integrate Indigenous perspectives in global marine governance [9]. Successful integration requires equitable partnerships, respect for sovereignty, and co-creation of knowledge, rather than one-way assimilation into Western science [10]. Challenges persist, including legal barriers, marginalization in decision-making, and threats from industrial development and climate change. Addressing these requires strengthening indigenous rights, investing in community capacity, and creating inclusive governance structures that honor traditional law and authority [11]. Ultimately, indigenous marine stewardship offers a transformative model of sustainability anchored in reciprocity, resilience, and respect for the ocean. Its integration with scientific knowledge is essential to build more just, effective, and lasting marine conservation solutions [1].

1.2 Co-management Practices

Co-management practices represent a collaborative governance model that bridges traditional knowledge and modern science to sustainably manage marine resources [12]. Involving shared authority and responsibility between governments and local communities, particularly co-management of indigenous and small-scale fishers, promotes inclusive, adaptive, and place-based stewardship of marine ecosystems [2]. Unlike top-down approaches, co-management is grounded in partnership, mutual respect, and power-sharing. It recognizes the value of local ecological knowledge (LEK) and Indigenous knowledge systems (IKS), integrating them with scientific research, policy frameworks, and monitoring tools to guide decision-making and management actions [13]. The core elements of effective co-management include joint decision-making, locally defined rules, conflict resolution mechanisms, and ongoing capacity building. Communities actively participate in setting harvest limits, establishing no-take zones, enforcing rules, and conducting resource assessments [14]. Government agencies, in turn, provide legal recognition, technical support, and access to scientific data [15]. Co-management enhances social legitimacy and compliance, as locally developed regulations are more culturally relevant and widely accepted [16]. It also improves ecological outcomes by leveraging the deep, place-specific knowledge of local stewards who observe environmental changes over time and adapt practices accordingly. Successful examples include Locally Managed Marine Areas (LMMAs) in the Pacific Islands, where customary marine tenure and traditional practices such as *tabu* are integrated with marine science and conservation planning [17]. In Canada and Australia, Indigenous Guardians and Sea Country programs blend traditional ecological knowledge with modern conservation science and monitoring. Despite its promise, co-management faces challenges such as unequal power dynamics, lack of long-term funding, legal barriers, and institutional resistance [18]. Building trust, ensuring legal recognition of community rights, and investing in inclusive capacity building are critical to overcoming these barriers. As climate change intensifies pressures on marine systems, co-management offers a resilient and adaptive governance model rooted in the local context and guided by cultural values and scientific evidence [14]. It fosters a shared sense of responsibility, strengthens conservation results, and empowers coastal and indigenous communities as rightful leaders in marine sustainability. In essence, co-management is a path to equitable and knowledge-rich marine governance, embodied in the future of integrated, sustainable ocean stewardship [19].

2 Marine Restoration and Climate Adaptation

2.1 Coral Restoration, Kelp Forests, and Mangrove Replanting

2.1.1 Coral Restoration

Coral restoration is a critical response to the accelerating decline of coral reefs due to climate change, ocean acidification, pollution, overfishing, and habitat destruction [20]. As vital ecosystems that support 25% of all marine species, provide coastal protection, and sustain millions of livelihoods, the restoration of coral reefs is essential for both biodiversity and human resilience [21]. Coral restoration involves active interventions to rehabilitate degraded reefs and enhance their ability to withstand future stress [22]. Core techniques include coral gardening, where fragments are grown in underwater or land nurseries before being transplanted back to damaged reefs, and microfragmentation, which accelerates coral growth by breaking colonies into tiny pieces that rapidly fuse and expand [23]. Another approach, larval enhancement, involves collecting coral spawn during natural reproduction events, rearing larvae in controlled settings, and then releasing them into degraded reef areas to promote natural settlement and genetic diversity [24]. This method is particularly valuable for restoring species with limited natural recruitment [25]. Assisted evolution and selective breeding are emerging strategies that aim to improve coral resilience by identifying and propagating genotypes that tolerate higher temperatures, changes in salinity, or acidification [26]. Coupled with probiotics and the manipulation of symbionts, these approaches seek to increase coral stress tolerance at both the cellular and ecosystem levels [27]. Coral restoration is increasingly integrated with community conservation, engaging local stakeholders, fishers, and tourism operators in planning, monitoring, and maintaining restored sites [28]. This fosters stewardship, creates alternative livelihoods, and improves local adaptive capacity [29]. However, restoration is not a substitute for broader conservation and climate action [30]. Its success is limited by addressing root causes of degradation, especially greenhouse gas emissions. Effective coral restoration must be part of a larger strategy that includes marine protected areas (MPAs), improvements in water quality, and climate mitigation [31]. The challenges remain to scale efforts, securing long-term funding and ensuring genetic diversity and ecological function in restored reefs [28]. Advances in restoration science, automation (e.g., coral planting robots), and global collaboration help overcome these barriers [32]. Ultimately, coral restoration offers hope for reef recovery and climate adaptation [33]. When guided by science, supported by policy and rooted in local partnerships, it serves as a powerful tool to rebuild the resilience of one of the most valuable and vulnerable ecosystems on Earth [33].

2.1.2 Kelp Forests

Kelp forests are highly productive marine ecosystems that provide critical habitat, food, and nursery grounds for countless marine species. They also provide ecosystem services such as carbon sequestration, shoreline protection, and support for coastal fisheries [34]. However, kelp forests around the world are in decline due to climate change, ocean warming, marine heatwaves, overgrazing by herbivores such as sea urchins, pollution, and habitat degradation [35]. The restoration and protection of kelp forests is now a growing focus within marine restoration and climate adaptation strategies. Kelp restoration involves both passive and active methods [36]. Passive restoration includes protecting natural kelp beds through marine protected areas (MPA), removing urchins, and reduction of pollution and sedimentation [37]. Active restoration involves direct interventions such as kelp transplantation, seeding with spores, and the deployment of artificial substrates to encourage kelp regrowth [38]. A key challenge in the recovery of kelp forests is urchin overgrazing, often due to the loss of natural predators, such as sea otters [39].

Restoration strategies often involve urchin culling or relocation of urchins, and in some areas, reintroduction of predator species to rebalance ecosystem dynamics [40]. Kelp aquaculture is also emerging as a dual solution for ecological restoration and sustainable economic development, offering carbon reduction and alternative livelihoods. The adaptation to climate change in kelp ecosystems focuses on enhancing resilience to warming and acidification [41]. Research into thermally tolerant kelp strains and selective breeding is underway to identify genotypes more resistant to rising sea temperatures. Monitoring technologies, such as remote sensing and underwater drones, help track forest health and guide restoration efforts [42]. Community and Indigenous participation is central to kelp restoration. In many coastal regions, kelp has cultural and subsistence value, and locally led initiatives have proven effective in restoring kelp beds while strengthening traditional stewardship practices [43]. Partnerships between scientists, local fishers, conservation groups, and governments are essential to long-term success. Despite their decline, kelp forests have great potential as nature-based solutions to climate adaptation, biodiversity conservation, and blue carbon strategies [44]. To realize this potential, restoration must be supported by robust science, adaptive policy, long-term investment, and global collaboration. Kelp forests represent one of the most valuable ecosystems restoring them is not only a path to ecological recovery, but a powerful strategy for building coastal resilience in a changing climate [45].

2.1.3 Mangrove Replanting

Mangrove replanting is a key strategy in marine restoration and climate adaptation, with the aim of reversing the loss of one of the most valuable coastal ecosystems [46]. Mangroves serve as critical buffers between land and sea, stabilizing shorelines, filtering pollutants, supporting fisheries, and providing nursery habitats for numerous marine species [47]. They are also highly effective carbon sinks, storing up to four times more carbon per hectare than terrestrial forests, making them vital to climate mitigation [48]. Despite their importance, mangroves have faced widespread degradation due to coastal development, aquaculture expansion, pollution, and deforestation [49]. Mangrove replanting initiatives are therefore essential to restore ecological function, protect coastal communities, and enhance biodiversity. Successful replanting involves more than simply planting trees [46]. It requires restoring hydrology, ensuring suitable tidal flow, sediment conditions, and salinity levels that allow natural regeneration. The most effective efforts prioritize ecological restoration focusing on conditions that support natural recolonization by native mangrove species over monoculture plantations that often fail in the long term [46]. Key techniques include direct propagule planting, seedling nurseries, and assisted natural regeneration [50]. Site selection is critical; planting must align with species-specific requirements and avoid high-energy shorelines or altered landscapes that are not suitable for root establishment [51]. Community involvement is a cornerstone of success, ensuring long-term stewardship, local knowledge integration, and shared economic benefits. Mangrove restoration provides benefits to climate adaptation by reducing coastal erosion, buffering storm surges, and protecting infrastructure [52]. In many regions, restored mangroves also support blue economy activities, such as sustainable fisheries, ecotourism, and honey production, creating incentives for conservation [53]. Global frameworks such as the UN Decade on Ecosystem Restoration and initiatives such as Mangroves for the Future and Blue Carbon partnerships are driving investment, policy support, and knowledge exchange for mangrove replanting [54]. However, challenges remain, including land tenure conflicts, lack of long-term monitoring, and projects that prioritize planting numbers over ecosystem function [55]. To ensure lasting impact, mangrove replanting must be scientifically based, community-driven, and integrated into broader coastal management and climate resilience strategies [56]. When done correctly, mangrove restoration is a powerful tool for protecting biodiversity, sequestering carbon, and building resilient coastlines in the face of a changing climate [46].

2.2 *Ocean Carbon Sequestration and Blue Carbon Ecosystems*

2.2.1 Ocean Carbon Sequestration

Ocean carbon sequestration is a vital climate adaptation and mitigation strategy that harnesses the natural ability to absorb and store atmospheric carbon dioxide (CO_2) [57]. As the largest carbon sink, the ocean currently absorbs approximately 25–30% of anthropogenic CO_2 emissions [58]. Enhancing this natural function is central to addressing climate change and supporting the resilience of marine ecosystems [6]. There are two main forms of ocean carbon sequestration: biological and geochemical. Biological sequestration involves the capture of carbon through the growth of marine organisms such as phytoplankton, seaweed, mangroves, salt marshes, and seagrasses collectively known as blue carbon ecosystems [57]. These systems fix CO_2 through photosynthesis and store it in biomass and sediments, sometimes for centuries. Restoring and protecting blue carbon habitats offers high-impact, nature-based solutions with co-benefits for biodiversity, fisheries, and coastal protection [59]. Emerging approaches also include macroalgae cultivation (e.g., kelp farming) for large-scale carbon uptake, with possibilities for harvesting and sinking biomass into deep waters. However, questions remain about long-term permanence, scalability, and ecological impacts [41]. Geochemical methods focus on enhancing ocean alkalinity to increase its capacity to absorb CO_2. This includes proposals such as adding crushed minerals (e.g., olivine) to seawater or electrochemical methods that increase carbonate ions [60]. These are still experimental and require careful evaluation of environmental risks and governance. Another proposed method is deep ocean storage, where captured CO_2 is injected into deep sea or sub-seafloor geological formations. While it may offer high storage capacity, concerns about leakage, monitoring challenges, and possible harm to deep-sea ecosystems [61]. Ocean carbon sequestration efforts must align with international climate frameworks and marine protection goals. Rigorous scientific research, transparent monitoring, and regulatory oversight are essential to ensure ecological integrity [62].

Ultimately, ocean carbon sequestration is not a silver bullet, but part of a broader portfolio of climate solutions. Its greatest potential lies in the natural, cost-effective, and community-inclusive approaches that simultaneously combat climate change and support marine life [63]. As climate impacts intensify, investing in ocean-based carbon solutions is a crucial step toward a more sustainable and resilient future [6].

2.2.2 Blue Carbon Ecosystems

Blue carbon ecosystems, mainly mangroves, seagrasses, and salt marshes, are coastal habitats that capture and store large amounts of atmospheric carbon dioxide (CO_2), making them critical allies in reducing climate change and marine restoration [64]. Although they cover only a small fraction of the ocean, these ecosystems sequester carbon at rates per unit area up to ten times higher than terrestrial forests, storing it in both plant biomass and deep anaerobic sediments where it can remain for centuries [57]. In addition to their carbon storage role, blue carbon ecosystems provide vital co-benefits: they support fisheries, protect coastlines from erosion and storm surges, filter pollutants, and provide habitat for diverse marine and bird species [65]. Their multifunctionality makes them a cornerstone of nature-based solutions for ecological restoration and community resilience [66]. However, blue carbon habitats are among the most threatened marine systems, facing degradation from coastal development, aquaculture, pollution, and rise [64]. When damaged, they not only lose their carbon capture function but also release stored carbon, exacerbating global emissions. Therefore, protecting and restoring these ecosystems is urgent and strategic [48]. Restoration strategies include replanting native vegetation, restoring tidal flow and sediment supply, and enabling natural regeneration [67]. Efforts must be site-specific, ecologically informed, and community-driven. Integrating traditional ecological knowledge enhances success, particularly in regions where local communities have managed coastal zones for generations [2]. Blue carbon is increasingly recognized in global climate policy, including under the Paris Agreement and Nationally Determined Contributions (NDCs) [68]. Projects are also emerging within carbon markets, where restored or protected blue carbon habitats generate verified carbon credits, which could unlock funding for conservation [54]. However, challenges remain with respect to carbon accounting, permanence, and equitable benefit sharing [69]. Scientific advances in remote sensing, sediment analysis, and ecosystem modeling are improving the accuracy of carbon assessments and inform better management [70]. However, governance frameworks must evolve to include blue carbon in coastal planning, climate adaptation strategies, and international carbon finance mechanisms [71]. In summary, blue carbon ecosystems represent a powerful, scalable, and cost-effective solution for climate mitigation, biodiversity conservation, and coastal resilience [72]. Investing in their protection and restoration is essential for a sustainable ocean future and for the achievement of global climate and development goals [73].

3 Education, Advocacy, and Public Engagement

3.1 *Increase Awareness and Behavior Change*

3.1.1 Increasing Awareness

Sensitivity raising is a critical component in the pursuit of marine ecology and ocean sustainability [74]. It involves informing, educating, and inspiring individuals, communities, and policymakers to recognize the importance of ocean health and the urgent need to address threats, such as pollution, overfishing, climate change, and biodiversity loss [75]. Without widespread public understanding and engagement, meaningful action at both the local and global levels remains limited [76]. Awareness begins with education, from early childhood through higher education and lifelong learning [77]. Integrating ocean literacy into school curriculum helps cultivate a generation that values marine ecosystems and understands their role in planetary health [78]. Interactive tools, citizen science programs, documentaries, and digital media platforms can expand this learning beyond classrooms, making marine science accessible and compelling to a broad audience [79]. Equally important is the strategic use of media and communication [80]. Storytelling, visuals, and data-driven narratives can translate complex scientific concepts into relatable messages that resonate emotionally and intellectually [81]. Social media platforms, influencers, and multimedia campaigns can rapidly amplify messages, reaching millions, and fostering global discourse on ocean issues [82]. Community participation plays a vital role in raising awareness. Localized campaigns that connect ocean sustainability to people's daily lives, such as sustainable seafood choices, plastic reduction, or beach cleanups, make abstract environmental issues tangible and actionable [83]. Culturally tailored outreach ensures inclusion and relevance, especially in coastal communities whose livelihoods are deeply tied to the sea [84]. Collaborations between scientists, educators, NGOs, artists, and the private sector can magnify awareness efforts [85]. It mobilizes behavioral change, empowers citizen advocacy, and builds the social and political will necessary for lasting solutions [86]. By embedding ocean awareness into our collective consciousness, we lay the foundation for a sustainable future where people and oceans can thrive together [7].

3.1.2 Behavior Change

Behavior change is a cornerstone of efforts to achieve marine ecology and ocean sustainability [87]. While policies, technologies, and conservation strategies are essential, their success ultimately depends on the willingness of individuals, communities, and institutions to adopt sustainable behaviors [88]. Changing behavior, be it reducing single-use plastics, choosing sustainable seafood, or supporting marine protection initiatives, requires a multifaceted approach grounded in education, social norms, and structural support [89]. At its core, behavior change is driven by awareness, motivation, and opportunity [90]. Education plays a fundamental role by providing the knowledge needed to understand the impact of daily actions on marine ecosystems [74]. However, information alone is not enough. Behavioral science shows that emotional engagement, social identity, and perceived efficacy strongly influence whether people act on what they know [91]. Campaigns that appeal to personal values, such as protecting children's health or preserving cultural heritage, tend to be more effective than those based solely on facts [92]. Social norms are powerful drivers. When sustainable behaviors become the expected standard, whether through peer influence, community leadership, or popular culture, individuals are more likely to conform [93].

Public commitments, social incentives, and visibility of positive actions (e.g., reusable products and clean-up participation) can normalize and reinforce ocean-friendly behaviors. Infrastructure and policy also play a critical enabling role [94]. For behavior change to be sustained, individuals need supportive environments. This includes access to alternatives (e.g., refill stations and recycling systems), economic incentives (e.g., subsidies for sustainable fisheries), and regulatory frameworks that discourage harmful practices [95]. Removing barriers and making sustainable choices easier, cheaper, or default option increases long-term adoption [96]. Additionally, behavior change is most effective when it is community-led and culturally relevant [97]. Local engagement, co-creation of solutions, and inclusion of traditional ecological knowledge foster ownership and relevance [98]. Digital tools, gamification, and behavior tracking apps can further personalize and encourage change. Ultimately, behavior change must not be viewed as a one-time shift but as a long-term cultural transformation [99]. Through consistent education, social reinforcement, and systemic support, sustainable behaviors can become ingrained, scalable, and resilient, ensuring healthier oceans for generations to come [100].

3.2 *Role of Media and Citizen Science*

3.2.1 Role of Media

The media plays a key role in shaping public perception, influencing behavior, and driving collective action toward marine ecology and ocean sustainability [101]. As a bridge between science and society, the media has the power to translate complex marine issues into accessible narratives that resonate with diverse audiences, mobilizing awareness and advocacy on a global scale [102]. Television, radio, newspapers, and documentaries in traditional media remain a powerful platform to educate the public about ocean health [103]. Compelling storytelling, supported by striking visuals and expert voices, can evoke emotional connections and urgency around issues, such as coral bleaching, plastic pollution, and the decline of marine species [104]. Award-winning nature documentaries and investigative journalism have often served as turning points in public discourse, influencing policy decisions and consumer behavior [105]. Digital and social media have further revolutionized the way marine issues are communicated. Platforms such as YouTube, Instagram, TikTok, and X (formerly Twitter) enable real-time content sharing, amplify grassroots movements, and democratize the creation and dissemination of information [106]. Influencers, scientists, NGOs, and citizen journalists now have the means to engage global audiences directly, building communities around marine conservation and sustainability [107]. The media also plays a watchdog role, holding governments and industries responsible by exposing illegal fishing, habitat destruction, or environmental negligence [108]. Investigative reporting can pressure policymakers and corporations to act, while sustained media coverage can keep ocean issues in the public eye [109]. Importantly, the media can frame marine sustainability not only as an environmental issue but also as a social, economic, and cultural concern [110]. By highlighting stories of coastal communities, indigenous knowledge, and innovative solutions, the media can present a more inclusive and hopeful narrative that motivates action across sectors [111]. Challenges remain, including misinformation, oversimplification, and media fatigue. To counteract these, partnerships between journalists, scientists, and educators are essential to ensure accurate, engaging, and impactful content [112]. Training programs in environmental communication can improve the capacity of the media to cover marine issues effectively. In essence, the media are not just a tool for information dissemination, but a powerful catalyst for change [113]. When strategically harnessed, it can increase ocean literacy, inspire behavioral changes, and drive the global momentum needed to secure a sustainable future for our oceans [114].

3.3 *Youth Engagement and Education Initiatives*

3.3.1 Engagement

Youth participation is a vital pathway to advance marine ecology and ocean sustainability, as young people represent both present stewards and future decision makers of our oceans [115]. Engaging young people promotes ocean literacy, environmental responsibility, and active participation in conservation, ensuring long-term social commitment to protecting marine ecosystems [116]. Young people are uniquely positioned to influence change through their energy, creativity, and adaptability [117]. Integrating ocean education into school curricula from primary to university levels lays the foundation for informed attitudes and behaviors [118]. Hands-on learning experiences such as field trips, citizen science projects, and marine internships improve understanding and personal connection to marine environments, making ocean conservation tangible rather than abstract [119]. Youth-led initiatives and organizations increasingly drive grassroots advocacy and innovation [120]. Through clubs, social media campaigns, and global movements such as Fridays for Future, young activists raise awareness, lobby for sustainable policies, and challenge harmful practices, such as plastic pollution and overfishing [121]. Their use of digital platforms amplifies their reach, mobilizing peers and communities around the world [122]. Empowering young people requires inclusive approaches that include diverse backgrounds, particularly involving marginalized and indigenous communities whose traditional knowledge can enrich ocean management [1]. Programs that support mentoring, leadership development, and capacity building equip young people with the skills and confidence to participate in scientific research, policymaking, and sustainable entrepreneurship [123]. Technology plays a crucial role in youth participation. Interactive applications, virtual reality experiences, and online learning platforms provide accessible tools for exploring marine science, fostering curiosity, and connecting young people worldwide [124]. These technologies can also facilitate participation in citizen science and advocacy campaigns, strengthening a sense of agency [125]. Furthermore, the participation of youth contributes to behavioral and cultural changes [126]. When young people adopt sustainable lifestyles and advocate for ocean-friendly policies, they influence family and community norms, creating ripple effects that extend beyond their immediate networks [127]. Ultimately, investing in youth engagement is an investment in the future of our oceans [128]. By nurturing young informed, passionate, and empowered leaders, society can build resilient, inclusive, and resilient movements that support marine ecosystems amid growing environmental challenges [129]. As youth voices gain greater recognition and support, they will be central drivers of innovation, conservation, and policy transformation toward a more sustainable ocean future.

3.3.2 Education Initiatives

Education initiatives are fundamental in advancing marine ecology and ocean sustainability, equipping individuals and communities with the knowledge, skills, and motivation necessary to protect and restore ocean health [87]. These initiatives span formal and informal learning, targeting diverse audiences, from schoolchildren to policymakers and the general public [130]. In formal education, the integration of ocean literacy into curricula at all levels is crucial [131]. Schools and universities increasingly adopt interdisciplinary approaches combining biology, ecology, social sciences, and policy to provide a holistic understanding of marine systems [132]. Experiential learning through fieldwork, marine laboratories, and internships deepens engagement by connecting theory with real-world ecosystems [133]. Programs such as ocean-themed science camps and marine research collaborations from universities increase student interest and prepare future marine professionals [134]. Informal education initiatives extend ocean learning beyond the classroom. Museums, aquariums, science centers, and public outreach events offer interactive exhibits and workshops that engage diverse audiences [135]. Community-based programs often involve citizen science, beach cleanups, and habitat restoration projects, promoting stewardship through hands-on participation [8]. These initiatives empower local communities, particularly coastal and indigenous populations, to apply traditional ecological knowledge along with scientific insights [136]. Digital technologies have transformed education initiatives, making marine science more accessible and scalable [137]. Online platforms, webinars, virtual reality experiences, and mobile apps deliver content around the world, breaking geographic and socioeconomic barriers [138]. Social media campaigns and storytelling amplify ocean messages, inspiring behavioral change and advocacy [139]. Collaborations between scientists, educators, NGOs, governments, and the private sector strengthen the reach and impact of education initiatives [140]. Such partnerships ensure content accuracy, cultural relevance, and alignment with conservation goals. Capacity building workshops and training for educators enhance their ability to teach complex marine topics effectively [141]. Evaluation and adaptive management are essential for improving education initiatives [142]. Monitoring participant knowledge gains, attitudes, and behaviors informs iterative refinement to maximize impact [143]. Inclusive strategies that prioritize gender equity, accessibility, and diversity improve participation and outcomes [118]. Ultimately, education initiatives cultivate an informed and engaged citizenry that values ocean sustainability [118]. By fostering curiosity, critical thinking, and responsibility, these efforts lay the foundation for sustained conservation actions, innovative solutions, and resilient marine ecosystems [144]. Education is not only a tool, but a catalyst that empowers societies to face the pressing challenges of marine ecology with knowledge and hope [78] (Figs. 4.1, 4.2 and 4.3).

Fig. 4.1 Students participating in coastal ecosystem education programs. Marine education programs promote environmental awareness and ocean literacy. **Source image**: Dr. Ferdinand GUENDEHOU (2026)

Fig. 4.2 Monitoring of coastal ecosystems. Environmental monitoring supports scientific data collection. **Source image**: Dr. Ferdinand GUENDEHOU (2026)

Fig. 4.3 Coral reef restoration through coral transplantation conducted by marine scientists to improve reef resilience and biodiversity. **Source:** Dr. Ferdinand GUENDEHOU in the photo above (2026)

References

1. Caldeira, M., Sekinairai, A. T., & Vierros, M. (2025). Weaving science and traditional knowledge: Toward sustainable solutions for ocean management. *Marine Policy, 174*, 106591.
2. Jennings, V., San Antonio, K. M., Brown, M. J., Choice, L., Simpson, Q., Ford, I., et al. (2024). Place-based conservation in coastal and marine ecosystems: The importance of engagement with underrepresented communities. *Sustainability, 16*(22), 9965.
3. West, S., Haider, L. J., Masterson, V., Enqvist, J. P., Svedin, U., & Tengö, M. (2018). Stewardship, care and relational values. *Current Opinion in Environmental Sustainability, 35*, 30–38.
4. Baharuddin, S. M. (2026). Living with the tides: Adaptive coastal knowledge and mangrove governance in Southeast Sulawesi. *Sustainable and Equitable Earth Development, 1*(1), 35–42.
5. Johnson, D. E., Parsons, M., & Fisher, K. (2022). Indigenous climate change adaptation: New directions for emerging scholarship. *Environment and Planning E: Nature and Space, 5*(3), 1541–1578.
6. Trebilco, R., Fleming, A., Hobday, A. J., Melbourne-Thomas, J., Meyer, A., McDonald, J., et al. (2022). Warming world, changing ocean: Mitigation and adaptation to support resilient marine systems. *Reviews in Fish Biology and Fisheries, 32*(1), 39–63.
7. Fischer, M., Maxwell, K., Nuunoq, Pedersen, H., Greeno, D., Jingwas, N., et al. (2022). Empowering her guardians to nurture our ocean's future. *Reviews in Fish Biology and Fisheries, 32*(1), 271–296.
8. Proulx, M., Ross, L., Macdonald, C., Fitzsimmons, S., & Smit, M. (2021). Indigenous traditional ecological knowledge and ocean observing: A review of successful partnerships. *Frontiers in Marine Science, 8*, 703938.
9. Huambachano, M., Soto, G. R. N., & Mwampamba, T. H. (2025). Making room for meaningful inclusion of indigenous and local knowledge in global assessments: Our experiences in the values assessment of the Intergovernmental Science-Policy Platform on Biodiversity and Ecosystem Services. *Ecology and Society, 30*(1).
10. Jones, T. J., Williams-Jones, G., & Nyce, H., Jr. (2025). Braiding indigenous knowledge systems and western science through co-creation and co-teaching. *Frontiers in Earth Science, 13*, 1587092.
11. Sakapaji, S. C., García Molinos, J., Parilova, V., Gavrilyeva, T., & Yakovleva, N. (2024). Navigating legal and regulatory frameworks to achieve the resilience and sustainability of indigenous socioecological systems. *Resources, 13*(4), 56.
12. Muhl, E. K., Armitage, D., Anderson, K., Boyko, C., Busilacchi, S., Butler, J., et al. (2023). Transitioning toward "deep" knowledge co-production in coastal and marine systems: Examining the interplay among governance, power, and knowledge. *Ecology and Society, 28*(4), 17.
13. Nishima-Miller, J., Johnson, L. R., Provencher, J. F., Alexander, S. M., Wilcox, A. A., Roy, C., et al. (2025). Bridging indigenous knowledge systems and western science for the co-management of wildlife in Canada: A systematic review. *Environmental Reviews, 33*, 1–19.
14. Choudhury, Y. (2025). *Comprehensive guide to fisheries co-management*. Educohack Press.
15. Arzberger, P., Schroeder, P., Beaulieu, A., Bowker, G., Casey, K., Laaksonen, L., et al. (2004). Promoting access to public research data for scientific, economic, and social development. *Data Science Journal, 3*, 135–152.
16. Sandström, A., Crona, B., & Bodin, Ö. (2014). Legitimacy in co-management: The impact of preexisting structures, social networks and governance strategies. *Environmental Policy and Governance, 24*(1), 60–76.
17. Love, M. W. (2021). Tabu, MPA and community: Nomenclature and the political ecology of marine conservation in Vanuatu. *Journal of Political Ecology, 28*(1), 91–110.

18. Nishima-Miller, J. (2025). *Transformative frameworks and strategic planning for wildlife stewardship, co-management, and indigenous-led conservation* (Doctoral dissertation, University of British Columbia).
19. dos Santos, P. C. (2020). Embedding sacred ecology in sustainability solutions: The role of post-secondary environmental science programs.
20. Good, A. M., & Bahr, K. D. (2021). The coral conservation crisis: Interacting local and global stressors reduce reef resiliency and create challenges for conservation solutions. *SN Applied Sciences, 3*(3), 312.
21. El-Naggar, H. A. (2020). Human impacts on coral reef ecosystem. In *Natural resources management and biological sciences*. IntechOpen.
22. Shaver, E. C., McLeod, E., Hein, M. Y., Palumbi, S. R., Quigley, K., Vardi, T., et al. (2022). A roadmap to integrating resilience into the practice of coral reef restoration. *Global Change Biology, 28*(16), 4751–4764.
23. Lennox, G. A. (2023). *Micro-fragmentation as a community-based coral restoration technique in a developing country* (Doctoral dissertation, Murdoch University).
24. Langley, C. (2025). *Optimising mass culturing of coral larvae from wild coral spawn: Enhancing survival and production for scalable reef restoration* (Doctoral dissertation, Southern Cross University).
25. Larson, J. E., Agneray, A. C., Boyd, C. S., Bradford, J. B., Kildisheva, O. A., Suding, K. N., & Copeland, S. M. (2023). A recruitment niche framework for improving seed-based restoration. *Restoration Ecology, 31*(7), e13959.
26. Van Oppen, M. J., Oliver, J. K., Putnam, H. M., & Gates, R. D. (2015). Building coral reef resilience through assisted evolution. *Proceedings of the National Academy of Sciences, 112*(8), 2307–2313.
27. Peixoto, R. S., Sweet, M., Villela, H. D., Cardoso, P., Thomas, T., Voolstra, C. R., et al. (2021). Coral probiotics: Premise, promise, prospects. *Annual Review of Animal Biosciences, 9*(1), 265–288.
28. Suggett, D. J., Edwards, M., Cotton, D., Hein, M., & Camp, E. F. (2023). An integrative framework for sustainable coral reef restoration. *One Earth, 6*(6), 666–681.
29. Cinner, J. E., Adger, W. N., Allison, E. H., Barnes, M. L., Brown, K., Cohen, P. J., et al. (2018). Building adaptive capacity to climate change in tropical coastal communities. *Nature Climate Change, 8*(2), 117–123.
30. Arneth, A., Olsson, L., Cowie, A., Erb, K. H., Hurlbert, M., Kurz, W. A., et al. (2021). Restoring degraded lands. *Annual Review of Environment and Resources, 46*(1), 569–599.
31. Filonchyk, M., Peterson, M. P., Zhang, L., Hurynovich, V., & He, Y. (2024). Greenhouse gases emissions and global climate change: Examining the influence of CO2, CH4, and N2O. *Science of the Total Environment, 935*, 173359.
32. Huang, W. (2025). Coral reef conservation: Challenges and advances in protecting coral reef ecosystems. *International Journal of Aquaculture, 15*.
33. Lockie, S., Paxton, G., Bartelet, H. A., Draaisma, L., & Brooksbank, J. L. (2025). Imagining outcomes of coral restoration and adaptation intervention in the Great Barrier Reef. *Restoration Ecology, 33*(8), e70173.
34. Cajete, G. A. (2020). Indigenous science, climate change, and indigenous community building: A framework of foundational perspectives for indigenous community resilience and revitalization. *Sustainability, 12*(22), 9569.
35. Edwards, M. S. (2025). Kelp forests under threat and the loss of ecosystem services they provide. *Annual Review of Ecology, Evolution, and Systematics, 56*, 365.
36. Eger, A. M., Marzinelli, E. M., Christie, H., Fagerli, C. W., Fujita, D., Gonzalez, A. P., et al. (2022). Global kelp forest restoration: Past lessons, present status, and future directions. *Biological Reviews, 97*(4), 1449–1475.
37. Bianchelli, S., Martini, F., Lo Martire, M., Danovaro, R., & Corinaldesi, C. (2023). Combining passive and active restoration to rehabilitate a historically polluted marine site. *Frontiers in Marine Science, 10*, 1213118.

38. Pinheiro, T. F. (2023). *Developing kelp reforestation methods* (Master's thesis, Universidade do Porto (Portugal)).
39. Wilman, E. A. (2021). Kelp forests: Catastrophes, resilience, and management. *Frontiers in Ecology and Evolution, 9*, 674792.
40. Miller, K. I., Blain, C. O., & Shears, N. T. (2022). Sea urchin removal as a tool for macroalgal restoration: A review on removing "the spiny enemies". *Frontiers in Marine Science, 9*, 831001.
41. Koch, S. (2025). *The sustainable expansion of seaweed as a restorative aquaculture: Using a carrying capacity framework that integrates barriers to expansion and ecosystem services* (Doctoral dissertation, Christian-Albrechts-Universität zu Kiel).
42. Diantari, R., Susanti, O., Wahid Yusuf, M., Yulianto, H., Wardiyanto, W., & Hasani, Q. (2018). Antibacterial activity of extract mangrove Avicennia sp., Rhizophora sp., and Bruguiera sp. against Vibrio harveyi and Vibrio parahaemolyticus.
43. Kesling, J. R., & McHugill, A. (2025). Restorative aquaculture stimulates cultural and ecological benefits: Centering sugar kelp, indigenous stewardship, and a New York estuary. *Earth Stewardship, 2*(2), e70012.
44. Thornton, T. F., & Scheer, A. M. (2012). Collaborative engagement of local and traditional knowledge and science in marine environments: A review. *Ecology and Society, 17*(3).
45. Aronson, J., Goodwin, N., Orlando, L., Eisenberg, C., & Cross, A. T. (2020). A world of possibilities: Six restoration strategies to support the United Nation's Decade on Ecosystem Restoration. *Restoration Ecology, 28*(4), 730–736.
46. Lovelock, C. E., Barbier, E., & Duarte, C. M. (2022). Tackling the mangrove restoration challenge. *PLoS Biology, 20*(10), e3001836.
47. Aransiola, S. A., et al. (2024). Niger Delta mangrove ecosystem: Biodiversity, past and present pollution, threat and mitigation. *Regional Studies in Marine Science, 75*, 103568.
48. Nunes, L. J., Meireles, C. I., Pinto Gomes, C. J., & Almeida Ribeiro, N. M. (2020). Forest contribution to climate change mitigation: Management oriented to carbon capture and storage. *Climate, 8*(2), 21.
49. Ferreira, A. C., Borges, R., & de Lacerda, L. D. (2022). Can sustainable development save mangroves. *Sustainability, 14*(3), 1263.
50. de Souza, D. C., & Engel, V. L. (2023). Advances, challenges, and directions for ecological restoration by direct seeding of trees: Lessons from Brazil. *Biological Conservation, 284*, 110172.
51. Braun, J. V., Robinson, R., Henry, A. L., & Kettenring, K. M. (2025). *Restore the shore: A guide to lakeshore revegetation in the Great Basin*. Utah State University Extension.
52. Hernández-Delgado, E. A. (2024). Coastal restoration challenges and strategies for small island developing states in the face of sea level rise and climate change. *Coasts, 4*(2), 235–286.
53. Guruge, T. P. S. R., & Kuruppu, I. V. (2025). Economically synchronized protection of coastal and marine ecosystems using an action impact matrix: Advancing a sustainable blue economy in South Asia. In *Handbook of sustainable blue economy* (pp. 1–26). Springer Nature Switzerland.
54. Li, P., & Liu, D. (2025). How blue carbon financing can sustain blue carbon ecosystems protection and restoration: A proposed conceptual framework for the blue carbon financing mechanism. *Ocean & Coastal Management, 265*, 107644.
55. Rakotonarivo, O. S., Rakotoarisoa, M., Rajaonarivelo, H. M., Raharijaona, S., Jones, J. P., & Hockley, N. (2023). Resolving land tenure security is essential to deliver forest restoration. *Communications Earth & Environment, 4*(1), 179.
56. Canevari-Luzardo, L. M., Mira, L., Williams, R., Lara, J., & Álvarez de Eulate, M. F. (2025). Building bridges: Leveraging science and livelihood development for community-driven mangrove restoration. In *Handbook of nature-based solutions to mitigation and adaptation to climate change* (pp. 1–31). Springer Nature Switzerland.

57. Khan, S., & Dhadse, S. (2025). Carbon sequestration by terrestrial and marine biodiversity—A tool for combating climate change. *Sustainability and Biodiversity Conservation, 4*(2), 69–88.
58. Gruber, N., Bakker, D. C., DeVries, T., Gregor, L., Hauck, J., Landschützer, P., et al. (2023). Trends and variability in the ocean carbon sink. *Nature Reviews Earth & Environment, 4*(2), 119–134.
59. Roberts, K. E., Rohr, T., Raven, M. R., Diamond, M. S., Visioni, D., Kravitz, B., et al. (2026). Potential impacts of climate interventions on marine ecosystems. *Reviews of Geophysics, 64*(1), e2024RG000876.
60. Eisaman, M., Geilert, S., Renforth, P., Bastianini, L., Campbell, J., Dale, A., et al. (2023). Assessing technical aspects of ocean alkalinity enhancement approaches. *State of the Planet Discussions, 2023*, 1–52.
61. Zendehboudi, S., Seyyedattar, M., Cata Saady, N. M., Mohammadzadeh, O., Mamudu, A., & Heagle, D. (2025). Exploring the potential of offshore geological CO2 storage in Canada: A comprehensive review and future outlook. *Energy & Fuels, 39*(13), 5987–6025.
62. Burns, W. (2023). Governance of ocean-based carbon dioxide removal research under the United Nations convention on the law of the sea. *Maine Law Review, 75*, 37.
63. Jena, S. R., Saha, S., & Agarwal, S. (2025). *Smart innovations for climate resilience, forest sustainability, and green economy*. Srjx Research and Innovation Lab Llp.
64. Alongi, D. M. (2023). Current status and emerging perspectives of coastal blue carbon ecosystems. *Carbon Footprints, 2*(3), 12.
65. Jahan, F. (2025). A systematic review of blue carbon potential in coastal marshlands: Opportunities for climate change mitigation and ecosystem resilience.
66. Di Pirro, E., Mendes, R., Fidélis, T., Sallustio, L., Roebeling, P., Marchetti, M., & Lasserre, B. (2022). The embeddedness of nature-based solutions in the recovery and resilience plans as multifunctional approaches to foster the climate transition: The cases of Italy and Portugal. *Land, 11*(8), 1254.
67. Wu, L., Ouyang, Y., Cai, L., Dai, J., & Wu, Y. (2023). Ecological restoration approaches for degraded muddy coasts: Recommendations and practice. *Ecological Indicators, 149*, 110182.
68. Sarun, H., & Hun, H. (2026). Blue carbon in nationally determined contributions (NDCs): Opportunities and challenges. *Journal of Agriculture and Environment, 3*(2), 225–230.
69. Craik, N. (2025). Equitable marine carbon dioxide removal: The legal basis for interstate benefit-sharing. *Climate Policy*, 1–15.
70. Dossa, K. F., & Miassi, Y. E. (2024). Remote sensing methods and GIS approaches for carbon sequestration measurement: A general review. *International Journal of Environment and Climate Change, 14*(7), 222–233.
71. Vanderklift, M. A., Herr, D., Lovelock, C. E., Murdiyarso, D., Raw, J. L., & Steven, A. D. (2022). A guide to international climate mitigation policy and finance frameworks relevant to the protection and restoration of blue carbon ecosystems. *Frontiers in Marine Science, 9*, 872064.
72. Wang, F., Liu, J., Qin, G., Zhang, J., Zhou, J., Wu, J., et al. (2023). Coastal blue carbon in China as a nature-based solution toward carbon neutrality. *The Innovation, 4*(5), 100481.
73. Waltham, N. J., Elliott, M., Lee, S. Y., Lovelock, C., Duarte, C. M., Buelow, C., et al. (2020). UN decade on ecosystem restoration 2021–2030—What chance for success in restoring coastal ecosystems? *Frontiers in Marine Science, 7*, 71.
74. Netto, M. P., Freire, P. L. S., Martins, G., Augusto, M., Abude, R. R. S., Cardoso, R. S., et al. (2025). A decade of ecological awareness on marine coastal ecosystems: University extension engaging basic education students in Brazil. *Marine Pollution Bulletin, 219*, 118315.
75. Laffoley, D., Baxter, J. M., Amon, D. J., Currie, D. E., Downs, C. A., Hall-Spencer, J. M., et al. (2020). Eight urgent, fundamental and simultaneous steps needed to restore ocean health, and the consequences for humanity and the planet of inaction or delay. *Aquatic Conservation: Marine and Freshwater Ecosystems, 30*(1), 194–208.

76. Khatibi, F. S., Dedekorkut-Howes, A., Howes, M., & Torabi, E. (2021). Can public awareness, knowledge and engagement improve climate change adaptation policies? *Discover Sustainability, 2*(1), 18.
77. Ramey, C. T., & Ramey, S. L. (2023). Early childhood education that promotes lifelong learning, health, and social well-being: The Abecedarian Project and its replications. *Medical Research Archives, 11*(11).
78. Bufalino, G. (2025). Blue education. Towards a new paradigm in environmental education. *Nuova Secondaria, 42*(6).
79. Şahin, E., Yücel, M., Uygurer, Z. P., & Salihoğlu, B. (2024). From onboard to the public: Science communication and media strategies from the R/V BİLİM-2 in addressing marine research on multi-stressors.
80. Verhoeven, P., Zerfass, A., Verčič, D., Moreno, Á., & Tench, R. (2020). Strategic communication across borders: Country and age effects in the practice of communication professionals in Europe. *International Journal of Strategic Communication, 14*(1), 60–72.
81. Gupta, N. N. (2023). Data-driven storytelling: How to use data to tell compelling stories and drive business outcomes. *World Journal of Advanced Engineering Technology and Sciences, 8*(1), 497–509.
82. Sharma, A., Gosai, H. G., & Singh, V. (2024). Role of global media and social media in raising environmental issues. In *Environmental activism and global media: Perspective from the past, present and future* (pp. 99–115). Springer Nature Switzerland.
83. Bujold, P. M., & Williamson, K. (2025). Solutions to plastic pollution with community-based behavior change.
84. Boateng, J. K., Golo, H. K., Ibrahim, S., & Erinosho, B. T. (2025). Examining the influence of Ghanaian coastal communities' cultural practices on the cultural rights of small-scale fishers: A focus on livelihood impacts. *Cogent Social Sciences, 11*(1), 2553222.
85. Lewandowski, M., & Czech, F. (2024). How universities and NGOs see each other: Review of literature on partnership concepts. *Universities and non-governmental organisations* (pp. 24–46).
86. Gautam, S., Gautam, A. S., Awasthi, A., & N, R. (2024). Citizen action and advocacy. In *Sustainable air: Strategies for cleaner atmosphere and healthier communities* (pp. 103–116). Springer Nature Switzerland.
87. Simanjuntak, M. B., Rafli, Z., & Utami, S. R. (2024). Enhancing maritime education for ocean sustainability: A multidisciplinary approach. *BIO Web of Conferences, 106*, 02006.
88. Altassan, A. (2023). Sustainable integration of solar energy, behavior change, and recycling practices in educational institutions: A holistic framework for environmental conservation and quality education. *Sustainability, 15*(20), 15157.
89. Fischbach, S., & Yauney, B. (2023). Social cognitive theory and reciprocal relationship: A guide to single-use plastic education for policymakers, business leaders and consumers. *Sustainability, 15*(5), 3946.
90. Michaelsen, M. M., & Esch, T. (2023). Understanding health behavior change by motivation and reward mechanisms: A review of the literature. *Frontiers in Behavioral Neuroscience, 17*, 1151918.
91. Hamann, K. R., Wullenkord, M. C., Reese, G., & Van Zomeren, M. (2024). Believing that we can change our world for the better: A Triple-A (Agent-Action-Aim) Framework of self-efficacy beliefs in the context of collective social and ecological aims. *Personality and Social Psychology Review, 28*(1), 11–53.
92. Summak, M. E., Erbay, M., & Algan Özkök, G. (2024). The role of public relations in promoting intangible cultural heritage. *Pakistan Journal of Life & Social Sciences, 22*(2).
93. Constantino, S. M., Sparkman, G., Kraft-Todd, G. T., Bicchieri, C., Centola, D., Shell-Duncan, B., et al. (2022). Scaling up change: A critical review and practical guide to harnessing social norms for climate action. *Psychological Science in the Public Interest, 23*(2), 50–97.

94. Rice, R. E., & Moxley, C. M. (2021). Plastic communication campaigns and interventions: Foci, theoretical frameworks, variables, and methods. In *The handbook of international trends in environmental communication* (pp. 132–154). Routledge.
95. Matthews, J. A., Matthews, S., Faries, M. D., & Wolever, R. Q. (2024). Supporting sustainable health behavior change: The whole is greater than the sum of its parts. *Mayo Clinic Proceedings: Innovations, Quality & Outcomes, 8*(3), 263–275.
96. Yue, C., et al. (2023). A behavioral approach to identify barriers to adoption of new technology: A case study of low-input turfgrasses. *Journal of Agricultural and Applied Economics, 55*(1), 72–99.
97. Meigs, R., et al. (2024). Community-led transformation principles: Transforming public health learning systems by centering authentic collaboration with community-based organizations. *Learning Health Systems, 8*(4), e10451.
98. Lang, W., Gkartzios, M., Yan, J., Chen, T., & Tan, S. (2024). Community co-creation through knowledge (co)production: The engagement of universities in promoting rural revitalization in China. *Journal of Rural Studies, 112*, 103455.
99. Woolley, K., & Sharif, M. A. (2026). Digital tracking, gamification, social media, and AI: How technology influences motivation. *Consumer Psychology Review, 9*(1), 85–100.
100. Del Mastro, A., Garofalo, G., & Veragouth, P. (2025). Terrestrial education and marine structural design: A post-sustainability framework for ocean engineering innovation. *Sustainable Marine Structures, 7*(3), 88–100.
101. Waters, Y. L., Wilson, K. A., & Dean, A. J. (2023). Plastic action or distraction? Marine plastic campaigns influence public engagement with climate change in both general and engaged audiences. *Marine Policy, 152*, 105580.
102. Muharram, M. (2025). Making the invisible visible: How the blue humanities translate climate change to the public. *Public Humanities, 1*, e120.
103. Otuya-Asohro, E. (2024). Role of the mass media in the sustainability of blue economy in Nigeria. *Future Sustainability, 2*(4), 1–7.
104. Gallagher, K. (2025). Reciprocity in human-ocean relationships: Towards the subversion of dominant environmental discourses.
105. Borins, S., & Herst, B. (2020). Beyond "Woodstein": Narratives of investigative journalism. *Journalism Practice, 14*(7), 769–790.
106. Singh, P., & Singh, A. K. (2025). *The digital revolution: Exploring new media in the 21st century*. Crown Publishing.
107. Jungmayr, P. J. (2024). *Sustainable social media: Consumer engagement with marine conservation NGOs* (Master's thesis, ISCTE-Instituto Universitario de Lisboa (Portugal)).
108. Katunzi, A., Ndati, N., & Thuo, J. (2025). Watchdog of natural resources transparency?: Evaluating print media oversight of Tanzania's extractive industry. *University of Dar es Salaam Library Journal, 20*(2), 101–118.
109. Bailey, I. (2022). Media coverage, attention cycles and the governance of plastics pollution. *Environmental Policy and Governance, 32*(5), 377–389.
110. Germond-Duret, C., & Germond, B. (2023). Media coverage of the blue economy in British newspapers: Sea blindness and sustainable development. *Geographical Journal, 189*(2), 193–203.
111. Porri, F., McConnachie, B., van der Walt, K. A., Wynberg, R., & Pattrick, P. (2023). Eco-creative nature-based solutions to transform urban coastlines, local coastal communities and enhance biodiversity through the lens of scientific and indigenous knowledge. *Cambridge Prisms: Coastal Futures, 1*, e17.
112. Purushothaman, G. (2025). Media literacy and public health communication: Managing misinformation and information overload during disease outbreaks in Nigeria. *SSRN 5747627.*
113. Mutinta, G. (2026). The role of the media in environmental communication. In *An African perspective on environmental communication: Advocating for sustainable environmental communication in Africa* (pp. 51–92). Springer Nature Switzerland.

114. Kelly, R., Francis, P., Shellock, R. J., Andrews, S., Arthur, B., Birkmanis, C. A., et al. (2025). Ocean literacy for ocean sustainability: Reflections from Australia. *Ocean and Society, 2*.
115. Kelly, R., Evans, K., Alexander, K., Bettiol, S., Corney, S., Cullen-Knox, C., et al. (2022). Connecting to the oceans: Supporting ocean literacy and public engagement. *Reviews in Fish Biology and Fisheries, 32*(1), 123–143.
116. Wang, L., et al. (2025). The promotive effect of ocean literacy on marine conservation behavior: A qualitative study based on Chinese university students. *PLoS One, 20*(8), e0323510.
117. Rahindra, H. A., & Wisnujati, N. S. (2024). Innovative empowerment strategies: Building a creative environment for young professionals in the small business sector. *Journal of Contemporary Administration and Management (ADMAN), 2*(1), 416–422.
118. Costa, R. L., & Faria, C. (2025). The blue school program: A model for holistic ocean literacy education. *Sustainability, 17*(2), 661.
119. Blanos, R., Beneventi, N., Peratoner, L., Peter Castelletto, L., & Petrera, F. (2023). Engaging with the public: The experience of MareDireFare a new science&art festival to promote the Ocean Decade. In *Book of Abstracts of the 1st International Conference on Ocean Education and Training* (pp. 51–52).
120. Purcell, W. M., Song, S., Witenstein, M. A., Tung, W. W., & Iyengar, R. (2025). Empowering youth-led innovation to accelerate sustainable development. *Journal of Education for Sustainable Development, 19*(2), 168–188.
121. Mancera, O. (2023). You don't need a sign to protest: The rise of digital climate activism. In *Handbook of multi-level climate actions* (pp. 150–170). Edward Elgar Publishing.
122. MacKinnon, K. R., Kia, H., & Lacombe-Duncan, A. (2021). Examining TikTok's potential for community-engaged digital knowledge mobilization with equity-seeking groups. *Journal of Medical Internet Research, 23*(12), e30315.
123. Chigudu, D. (2025). Empowering African youth for sustainable peace: The role of skills training and entrepreneurship. *International Journal of Research in Business & Social Science, 14*(5), 548.
124. Iburahim, S. A., Naidu, B. C., & Ananthan, P. S. (2025). Virtual reality and augmented reality. In *Information technology in fisheries and aquaculture* (p. 109). Springer Singapore.
125. Anyolitho, M. K., Huyse, T., Masquillier, C., Nyakato, V. N., & Poels, K. (2024). Empowering communities through citizen science and participatory action research: Implementation of a schistosomiasis communication campaign in Uganda. *Humanities & Social Sciences Communications, 11*(1), 1–11.
126. Singh, R. (2025). Youth and cultural shift: Impact on social responsibilities. *Idealistic Journal of Advanced Research in Progressive Spectrums (IJARPS), 4*(3), 98–101. *eISSN–2583-6986*.
127. Muktiarni, M. M. M., Widiaty, I., Widaningsih, L., & Yulia, C. (2025). Framing ocean literacy as a policy tool to address marine litter within sustainability transitions. *Journal of Ship and Marine Structures, 10*(1), 25–43.
128. Kurniawan, M. D., & Do, C. V. (2025). Enabling youth participation in the blue economy: A comparative analysis of policies and legal frameworks in the EU and ASEAN. In *Handbook of sustainable blue economy* (pp. 1–32). Springer Nature Switzerland.
129. Fernandez, D. (2025). 18. Advancing youth-led movements for the ocean is essential. In *Navigating our way to solutions in marine conservation* (p. 277). Open Book Publishers.
130. Mutanda, G. W., & Nhamo, G. (2025). Ocean and marine literacy in formal education: Systematic review informing engagements with UN ocean decade. *Environmental Education Research, 32*, 1–35.
131. Alexander, K. A., Hobday, A. J., Cvitanovic, C., Ogier, E., Nash, K. L., Cottrell, R. S., et al. (2018). Progress in integrating natural and social science in marine ecosystem-based management research. *Marine and Freshwater Research, 70*(1), 71–83.
132. Moody, H. (2025). Applied scientific research and action in experiential learning: The Indian River Lagoon, Florida. In *Experiential learning in geography: The world as our classroom* (pp. 91–104). Springer Nature Switzerland.

133. Mauro, C. (2025). Accessible marine science: Authentic lessons and real-world stewardship at the Navarre Beach Marine Science Station. *Current: The Journal of Marine Education, 40*(1).
134. Erickson, C. (2021). *The nature of informal educators in science centers, science museums and zoos: Case studies of personnel, practices, programs, outreach and organization*. University of Arkansas.
135. Rossi, S. (2022). Science monitoring and scientific outreach. In *SDG 14: Life below water: A machine-generated overview of recent literature* (pp. 535–596). Springer International Publishing AG.
136. Shahinyan, G., & Kazanjian, G. (2025). Surfing knowledge: The Iliad Academy to support ocean literacy and digitization efforts. In *OCEANS 2025 Brest* (pp. 1–5). IEEE.
137. Yandrapalli, V. K., & Sharma, S. (2025). Cloud-powered distance learning: Breaking geographical barriers in education. In *Cloud computing for smart education and collaborative learning* (pp. 133–144). Chapman and Hall/CRC.
138. Jain, A. (2025). Sustainable storytelling engaging audiences with eco. In *Driving business success through eco-friendly strategies* (p. 441). IGI Global.
139. Batley, R., & Rose, P. (2013). Collaboration in delivering education: Relations between governments and NGOs in South Asia. In *Achieving education for all through public–private partnerships?* (pp. 107–113). Routledge.
140. Rahabav, P., Kempa, R., Tomasila, B., Singerin, S., Sohilait, D., Rumfot, S., et al. (2025). Empowering coastal communities for sustainable marine conservation: Integrating education, technology, and local engagement in Negeri Mamala. *Devotionis*, 33–40.
141. Koh, G. A., & Askell-Williams, H. (2021). Sustainable school-improvement in complex adaptive systems: A scoping review. *Review of Education, 9*(1), 281–314.
142. Sucala, M., Ezeanochie, N. P., Cole-Lewis, H., & Turgiss, J. (2020). An iterative, interdisciplinary, collaborative framework for developing and evaluating digital behavior change interventions. *Translational Behavioral Medicine, 10*(6), 1538–1548.
143. Ezekwu, E. (2025). Empowering girls through inclusive education systems focused on equity and cultural diversity. *International Journal of Research Publication and Reviews, 6*(1), 1839–1854.
144. Isangadighi, G., & Ukudo, B. (2025). Perceptions and awareness of air quality and its health impacts in Agbarho community, Ughelli North, Delta State. *Journal of African Innovation and Advanced Studies, 7*.

Index

F. J. C. Guendehou, K. L. Ayitchéhou, *Marine Ecology and Ocean Sustainability*, https://doi.org/10.1007/978-3-032-32756-7

The manufacturer's authorised representative in the EU is Springer Nature Customer Service Centre GmbH, Europaplatz 3, 69115 Heidelberg, Germany. If you have any concerns regarding our products, please contact ProductSafety@springernature.com

Printed and bound by CPI Group (UK) Ltd, Croydon, CR0 4YY
17/09/2026
02218451-0001